THE VIOLENT UNIVERSE

THE JOHNS HOPKINS UNIVERSITY PRESS   *Baltimore and London*

THE VIOLENT UNIVERSE

Joyrides through the X-ray Cosmos

KIMBERLY WEAVER

Foreword by RICCARDO GIACCONI

© 2005 The Johns Hopkins University Press
All rights reserved. Published 2005
Printed in the United States of America on acid-free
paper
9 8 7 6 5 4 3 2 1

The Johns Hopkins University Press
2715 North Charles Street
Baltimore, Maryland 21218-4363
www.press.jhu.edu

Library of Congress Cataloging-in-Publication Data

Weaver, Kimberly, 1964–
    The violent universe: joyrides through the X-ray
cosmos / Kimberly Weaver; foreword by Riccardo
Giacconi
        p. cm.
Includes index.
ISBN 0-8018-8115-3 (hardcover: alk. paper)
1. X-ray astronomy. I. Title
    QB472.W43 2005
522'.6863—dc22   2004021145

A catalog record for this book is available from the
British Library.

Contents

Foreword

I N AN ATTRACTIVE and comprehensive form, this book brings to the general reader views of the cosmos as seen with x-ray eyes. The universe revealed by x-rays is one pervaded by cosmic explosions and the infall of matter in extremely massive or collapsed objects. In such events, matter is heated to millions of degrees, creating vast expanses of hot plasma and accelerating particles to nearly the speed of light. The characteristic radiation emitted by this plasma, as well as by the interaction of high-energy particles with fields or matter, is in the x-ray range of wavelengths.

X-ray observations have led to the discovery of previously unsuspected physical processes and states of matter that would not have been possible through visible light observations. In the past forty years, x-ray astronomy has uncovered the existence of binary stellar systems containing neutron stars and black holes of stellar mass, giving a new insight into stellar evolution. And it has shown that most of the universe's normal matter is contained in multimillion-degree plasmas that pervade the space between galaxies and are a major component of the cosmic web. Because of x-ray observations, we now know that high-energy cosmic events play a fundamental role in the formation and dynamic evolution of structures in the universe.

The incredibly rapid development of x-ray instrumentation, and x-ray telescopes in particular, has made these discoveries possible. Over only four decades, the sensitivity of the observations has increased a billion-fold. In visible light astronomy, it took four hundred years to move from the naked eye to the Hubble Telescope. The x-ray telescope at the Chandra Observatory is able to provide pictures of the sky with an angular resolution comparable to those produced by visible light observatories on the ground.

Within these pages, Kimberly Weaver has done an excellent job of compiling a selection of images that demonstrate both the scientific value offered by x-ray observation and the sheer beauty of the cosmos. By comparing x-ray images with those made with visible, infrared, and radio telescopes, she shows how modern astronomers use the all-wavelength approach to the study of space, which offers the best hope to advance our understanding of the mysteries of dark matter, dark energy, and of the origin and evolution of the universe. Weaver has made it possible for everyone to share in this great intellectual adventure.

RICCARDO GIACCONI
The Johns Hopkins University

Preface

M Y LOVE OF ASTRONOMY began when I was a small child. I was fascinated with a series of science books that my father had purchased from the LIFE Nature Library, and I would often sneak away into a quiet corner of the house and read them from cover to cover. My favorite book from the series was called *The Universe,* which contained all sorts of artists' impressions and grainy optical images of comets, planets, stars, and galaxies. I distinctly remember knowing that the photographs were fuzzy, and I wondered what these objects would look like if I could view them close up. I am thrilled to say that I have finally had that opportunity. It is so amazing to see the beautiful images from the Hubble Space Telescope and realize just how much detail my favorite photographs in *The Universe* were missing.

When the images of the 1960s taken with ground-based optical telescopes are compared to today's Hubble images, they obviously pale in comparison. But what most people don't know is that our technology for detecting other types of light from celestial objects, such as x-rays and gamma rays, has also improved dramatically over the past few decades. Advances in our knowledge of these fields have resulted in rapid leaps in our understanding of the universe. Equivalent progress took hundreds of years in the field of optical astronomy. Still, it is a challenge to explain our new views of the heavens, because people are not used to learning about objects that they can't see. For example, even though several pages of *The Universe* were devoted to discussing radio telescopes and radio astronomy, only a single small drawing of a radio picture appeared in the book. The omission suggests to me that when the book was published in 1962, the editors felt that the general public might not understand radio pictures.

A reluctance to emphasize nonoptical astronomy is not surprising. When talking to people about astronomy, I frequently encounter confusion about my particular field. I think this confusion exists because we astronomers have been very poor communicators. When people ask what I do and I tell them that I am an x-ray astronomer, they often react with raised eyebrows and a somewhat puzzled look. They know that astronomers use telescopes to observe the heavens, but the word *x-ray* doesn't seem to fit. Their next question is usually, "What's x-ray astronomy?" I have learned that the best way to explain x-ray astronomy to my family and friends is to show them an x-ray picture. Pictures are something everyone can relate to, and today's x-ray pictures of our universe can be just as spectacular and informative as pictures obtained in visible light.

I decided to write this book because I want to tell the story of x-ray astronomy. I also want to share my fascination with x-ray astronomy with the widest audience possible, but especially with my family and friends. In only forty years, the science of x-ray astronomy has given us an entirely new way of looking at the universe. The beautiful images in these pages from extraordinary satellites like the Chandra X-ray Observatory (Chandra) and the X-ray Multi-Mirror Newton Observatory (XMM-Newton) can awe and inspire us in many ways. My hope is that this book will help to show that there's more to the universe than meets the eye!

The idea for a picture book of x-ray images geared to the nonastronomer came to me in 1995 while I was working at the Johns Hopkins University. There are several people I would like to acknowledge who have helped me in many ways during this project. First, I want to thank Julian Krolik, who strongly encouraged me to pursue this book and who provided much motivation over the years. I would also like to thank Nancy Levenson and Tahir Yaqoob, who discussed this book with me on several occasions. Of course, this book would not have been possible without the Chandra X-ray Observatory. I would like to express my gratitude to Martin Weisskopf, the Chandra project scientist, Harvey Tananbaum, the director of the Chandra X-ray Center, and the entire Chandra team and the National Aeronautics and Space Administration for building, launching, and operating such a remarkable satellite. I am also grateful to the talented technical and artistic staff of the Smithsonian Astrophysical Observatory Chandra X-ray Center, who created many of the exquisite images in this book and made them easily available to the public on a well-designed and attractive website. I wish to express my appreciation to the scientists who produced many of the images in this book, especially Andrew Wilson, David Batchelor, Namir Kassim, and Philip Blanco, who offered special help. I would also like to thank my best friend and companion, Jack Scheer, for providing support and encouragement to continue with this book and for reading the draft manuscript. Finally, I want to express my appreciation and gratitude to the National Aeronautics and Space Administration of the 1960s and 1970s, which inspired me to become a scientist and astronomer. May the inspiration continue.

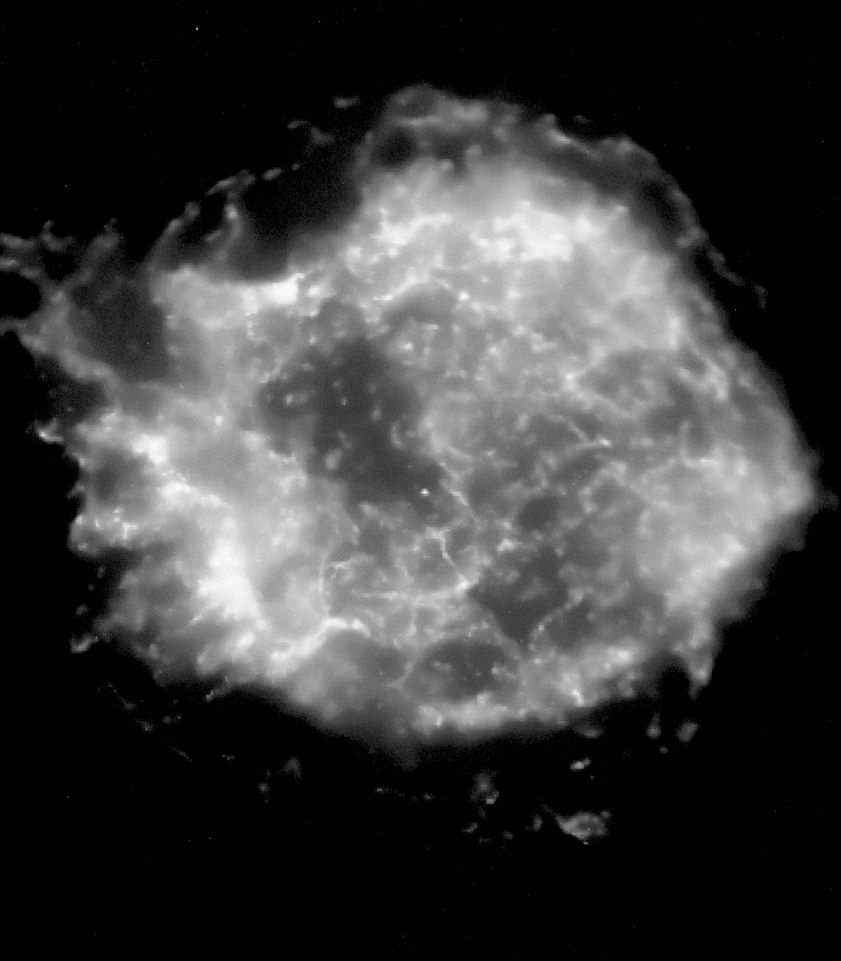

The Story So Far

OUR UNIVERSE is explosive and violent. The ominous black holes, plasma jets, stellar explosions, and cosmic collisions that have become familiar thanks to modern science fiction are not merely the notions of storytellers. They are real galactic events marked by excessive temperatures, extreme densities, intense magnetic fields, and crushing gravity. Such unusual physical conditions do not exist naturally on Earth, and we cannot reproduce them in our laboratories. So how can we know what they are like? Fortunately, these extreme and violent places shine brightly—not necessarily in visible light, but in x-rays. These x-rays, just like those used in medicine, are high-energy particles that travel at the speed of light. In recent years, scientists have learned to harness them and mine them for information. This is the exciting field of x-ray astronomy.

Astronomy is the study of natural objects that exist outside of the Earth's atmosphere. In many ways, it is the oldest science. The roots of astronomy can be traced to ancient times, with the earliest recorded observations dating from around 2000 BC. The ancient cultures of Egypt, Britain, Central America, and China were especially fascinated by astronomy. We witness their fascination in the Egyptian pyramids, Stonehenge, Mayan temples, and ancient Chinese palaces, which were frequently aligned with the positions of the Sun and Moon, serving as solar and lunar observatories. People in early cultures watched the movements of the Sun, Moon, stars, and planets with great curiosity and realized that the motions corresponded to regular terrestrial events. By learning these cycles, it was possible to predict the changing seasons, the rising tides, the best time to plant crops, and other aspects of nature that were important for their survival. Ancient astronomers created calendars based on the solar and lunar cycles. But people didn't know what made celestial objects move in the sky, and so these apparently life-giving and life-saving objects were sometimes worshiped as gods.

The ancient Greeks carried out the first serious investigations of the universe, combining observations, mathematics, and philosophy with a powerful belief in perfection. When it was possible to perform simple observations to test an idea, their investigations often led to the right answer. Philosophers such as Plato (427–347 BC) understood that the Earth was round. The Greeks were aware of the circular shadow cast by the Earth on the Moon during a lunar eclipse and had also noticed that the mast of a ship appeared before the hull as the ship approached on a curved horizon. On the other hand, when simple observations were not enough to test an idea, mere reasoning did not necessarily provide the right answer. For example, the Greeks' observations of the stars and planets were misleading. To them, the stars in the night sky appeared as points of light that traveled across the sky while the Earth remained stationary. They did not realize that the Earth rotates on its axis, causing the stars to rise and set from our perspective. So Plato and later Aristotle believed that the Earth was located at the center of the cosmos, and that the Sun, Moon, planets, and stars revolved around the Earth.

The ancient Greeks' belief in an Earth-centered universe began a debate that would last for centuries. Around 280 BC, Aristarchus of Samos estimated the distances from the Earth to the Sun and from the Earth to the Moon. He correctly deduced that because the Sun was so much farther away than the Moon, it had to be much larger than the Moon. Knowing the relative sizes of the Earth, Sun, and Moon allowed Aristarchus to reason that the Earth and Moon must revolve around the Sun.

Unfortunately, this idea was dismissed at the time because the accepted belief was that the Earth lay at the center of the cosmos. But there were serious problems with the Earth-centered view, such as the fact that planets sometimes appeared to move backwards in the sky due to their orbits around the Sun. Over the next few hundred years, the astronomer Hipparchus, and later Ptolemy, attempted to explain the curious backward motion of the planets using mathematics, contending that the planets moved in a complex series of circles. Although incorrect, their explanation was accepted because it agreed with the Earth-centered view. It wasn't until the sixteenth century that Polish-born astronomer Nicolaus Copernicus revived Aristarchus's idea that the Earth orbits the Sun, correctly deducing that the Sun is the center of the entire solar system.

Our concept of the heavens advanced dramatically in 1609–10, when Galileo Galilei and his contemporaries turned their newly invented optical telescopes toward the night sky and gathered the first detailed astronomical observations. The power of telescopes to magnify celestial objects allowed people to see the true nature of the planets for the first time. Galileo's data on the movement of the tiny moons of Jupiter contradicted the idea of an Earth-centered universe, because the moons were clearly not orbiting the Earth. Galileo also discovered that Venus had phases, like Earth's Moon. The fully lighted phase when Venus is near the Sun could only be explained if Venus were orbiting the Sun. With these improved observations, the ancient Greeks' belief that the Earth was the center of the solar system was at last completely debunked, and Aristarchus and Copernicus were proven correct. In a separate investigation, using precise visual records of the planets by the Danish astronomer Tycho Brahe, Johannes Kepler showed that planets move around the Sun in ellipses, not in circles, as Hipparchus and Ptolemy had theorized. Later, Isaac Newton devised a law of gravitation that astronomers used for almost 300 years, until Einstein's general theory of relativity superseded it.

THE ELECTROMAGNETIC SPECTRUM

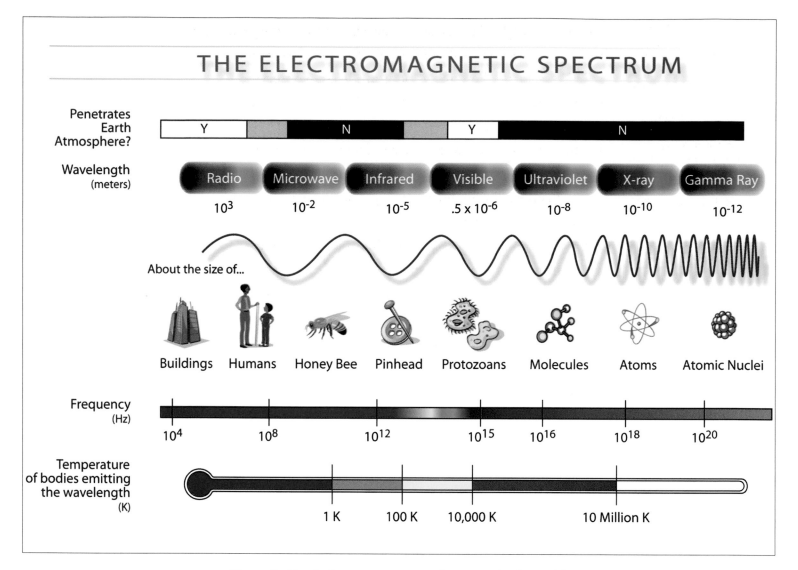

Since Galileo's day, astronomers have worked steadily to improve the resolving and collecting powers of their telescopes. In less than 400 years, the sensitivity of optical telescopes has improved by 100 million times. Bigger telescopes were built and placed on mountaintops, where the atmosphere is thinner and there is less distortion from the atmosphere. The first astronomical photographs were recorded in the 1840s, allowing astronomers to collect large amounts of light and to keep detailed records of their observations. As the sensitivity of film has improved, and now with digital technology, astronomers have been able to resolve fainter objects and peer farther into space. Astronomers have also waged war against light pollution

FIGURE 1.1. Light. This illustration of the electromagnetic spectrum shows the relationship
between different types of light,
from radio waves (longest wavelength, lowest frequency) to
gamma rays (shortest wavelength, highest frequency).
Objects that produce long-
wavelength, low-frequency light
have the lowest temperatures,
while objects that produce short-
wavelength, high-frequency light
have the highest temperatures.
The drawings of various recognizable objects, not all of which
give off light themselves, illustrate the relative sizes of wavelengths of light from across the
spectrum. The electromagnetic
spectrum is discussed in more
detail in Chapter 2.

to keep the night skies as dark as possible, so that the glow from
our cities does not outshine the light from faint stars and galaxies. To escape the limitations of ground-based observing
entirely, the National Aeronautics and Space Administration
(NASA) launched the Hubble Space Telescope into space in
1990. This telescope is the source of many of today's most beautiful and detailed visible images of the universe.

Visible light provides only a partial view of the universe,
however, because celestial objects produce radiation at all wavelengths of the electromagnetic (light) spectrum. Figure 1.1 shows
an illustration of the entire spectrum of light, from radio waves
to gamma rays. In a world where we are so used to relying on
our eyes for information, the thought of "seeing" something that
is invisible is puzzling at first. But the ability to perceive the
invisible is really not so strange if you think about it. For example, we don't see sound waves, but our ears are perfect sound
wave detectors. On the day that Galileo first looked through his
telescope, he was seeing only a fraction of the electromagnetic
spectrum. It would take 200 years for a new form of light to be
perceived.

Hold your hand over an electric stove. Before the rings
glow red, you can feel heat. Your hand is an infrared detector. It
was infrared—the means by which your TV remote works—that
next became available to astronomers. Sir William Herschel, a
German-born scientist, who is also known for discovering the
planet Uranus, discovered infrared light in 1800. Herschel used
a prism to divide sunlight into a rainbow of colors (red, orange,
yellow, green, blue, and violet) because he wanted to see if the
different colors of light had different temperatures. By placing a
thermometer in each colored patch of light, he found that the
colors became progressively warmer, from violet to red. Then he
placed a thermometer just beyond the red part of the rainbow,
in a region where there appeared to be no sunlight, and found
that this spot was the warmest. (The result of Herschel's experiment appears to contradict the fact that shorter wavelengths of

The Story So Far

light correspond to higher temperatures. However, this is because the prism spreads out blue light more than red light. The energy in the red portion of the spectrum is more concentrated, causing the thermometer to register a higher temperature.) Herschel was the first to realize that heat, or thermal radiation, is, in fact, a form of light that we cannot see with our eyes—infrared light.

One year after the discovery of infrared light, the chemist Johann Wilhelm Ritter followed Herschel's lead and looked for invisible light beyond the other end of the visible spectrum, where he found what we now call ultraviolet light. Ultraviolet light is what causes your skin to burn if you spend too much time in the sun. Ritter used a colorless chemical containing silver that was known to decompose and turn black when exposed to light, similar to what happens when photographic film is exposed to light. Just as Herschel had used a thermometer to measure changes in heat, Ritter exposed his chemical to different colors of light and found that it blackened faster when exposed to violet light than to red light. Surprisingly, the fastest reaction occurred beyond the violet region, where no light could be seen. This marked the discovery of ultraviolet light.

The radio portion of the electromagnetic spectrum was discovered next, in 1885. Today, radio waves carry all sorts of information through the atmosphere, between antennas on the ground and up to satellites in space, bringing music to your radio, images to your television, and allowing you to have convenient conversations on your cell phone. Man-made radio transmissions were first sent and received by the German physicist Heinrich Hertz, for whom the unit of frequency, the hertz (Hz), is named. Hertz produced radio waves in his laboratory and showed that they could be reflected in the same way as light. The first celestial source of radio waves was found by accident in 1931 when Karl Jansky, a radio engineer, was using a radio antenna to search for the sources of static in long-distance telephone calls. He eventually found that the noise was coming from the center of our galaxy in the constellation Sagittarius. A few years later, in 1937, the first radio telescope designed solely to study celestial sources was built, and radio astronomy quickly grew into a powerful means of observing the heavens.

X-rays were discovered ten years after radio waves. In 1895, the German physicist Wilhelm Roentgen was experimenting with cathode ray tubes, which are similar to today's fluorescent light bulbs. They are filled with a special gas and they glow when a high electric current is passed through them. In his laboratory, Roentgen had covered a cathode ray tube with heavy black paper so that no light could escape. When he charged the tube, he noticed to his surprise that an object with a metallic coating sitting across the room began to glow. This "escape" of some sort of invisible light ray that could penetrate dark paper and interact with the metal was an amazing discovery. Roentgen called these rays x-rays. When testing these rays, he found that if he held his hand in front of the tube, he could see the bones. This demonstration that x-rays can pass right through skin and muscle but not bone was the first hint that they could be used for medical purposes.

Gamma rays were discovered next, in 1900, by Paul Villard, a French physicist with an interest in chemistry. Gamma rays are the strongest form of electromagnetic radiation, and they are produced when the nucleus of an atom breaks down and releases its energy. We don't have much experience with gamma rays in everyday life, which is fortunate, because they can be lethal. While experimenting with x-rays and radioactive elements, Villard discovered this new type of radiation that came from the radioactive elements. He knew that these rays were different from x-rays, because they could penetrate much deeper into matter than x-rays could.

The microwave portion of the electromagnetic spectrum was discovered most recently. Microwaves, the same form of energy that pops our popcorn, were first generated by experiment in 1932, but just like the discovery of radio waves from

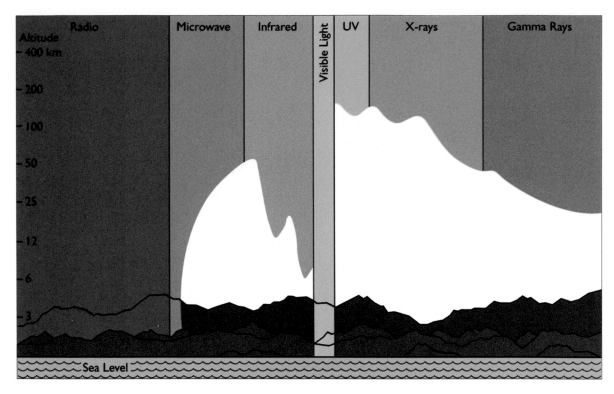

FIGURE I.2. Atmospheric windows. This illustration shows the transparency of Earth's atmosphere to different parts of the electromagnetic spectrum. There are three main "windows" in the atmosphere for detecting radio, microwave, and visible light. It is also possible to detect infrared light at high altitudes. But the atmosphere blocks ultraviolet, x-ray, and gamma-ray light, so we can only observe these from space.

space, the first celestial source of microwave radiation was found by accident. In the early 1960s, Arno Penzias and Robert Wilson of Bell Telephone Laboratories were detecting a source of excess noise in a sensitive antenna they were using to conduct a radio astronomy experiment. They eventually realized that this was the cosmic microwave background radiation, which is believed to be the heat left over from the Big Bang that created our universe. If we could see this microwave light unaided, the sky would glow brightly in every direction (see Chapter 8).

Astronomical objects produce all of these forms of electromagnetic radiation, but the Earth's atmosphere is transparent only to radio and visible light (fig. 1.2), and to a small portion of the microwave and infrared spectrum. Radiation with energies greater than visible light will interact with the atoms in the Earth's atmosphere and be scattered away. Because of the limiting effects of the atmosphere, observations of the most energetic forms of light, such as ultraviolet light, x-rays, and gamma rays had to wait until we could place detectors in orbit above the Earth's atmosphere. This meant that the high-energy universe remained invisible to astronomers until the space age, which dawned with the launching of rockets in the 1940s and began in earnest with the launching of the first artificial satellites in the 1950s.

The field of x-ray astronomy is only about 40 years old, but in that time, our x-ray telescopes have evolved as much as optical telescopes did in the course of 400 years. Even in its infancy, x-ray astronomy has discovered fascinating new phenomena, such as the black holes predicted by Einstein's theories. The next chapter explores x-ray astronomy in more detail, how x-ray light compares with other types of light, how x-rays are detected, how x-ray images are created, what types of celestial objects produce x-rays, and what x-rays can tell us about our universe. ✳

2

What Our Eyes Can't See

L IGHT IS A TYPE of energy that travels from one place to another. A handy way to think of light is that it has multiple personalities—sometimes it behaves like a collection of particles, called photons, and sometimes it behaves like a wave. One form of a wave is the up-and-down motion of water that you feel if you are in a boat on a lake and another boat passes by. A wave can be defined by its wavelength, which is the distance between two crests of the wave, or by its frequency, which is the number of crests that pass a certain point each second. Wavelength is the trait that makes one type of light different from another.

Our Sun's glow is primarily comprised of visible light. This light is made up of all of the colors of the rainbow—red, orange, yellow, green, blue, and violet. Every color has its own wavelength and frequency. Red light has a longer wavelength than green light, which has a longer wavelength than violet light. The wavelength of light is also related to its energy. Longer waves (lower frequencies) carry less energy than shorter waves (higher frequencies), and so violet light is more energetic than red light (fig. 1.1).

The electromagnetic spectrum is made up of light of all wavelengths, frequencies, and energies, from radio waves, the weakest radiation, to gamma rays, the strongest radiation. Gamma rays are the tiniest waves—about the size of an atomic nucleus—and they carry hundreds of thousands to a million times more energy than a visible photon. X-rays are larger than

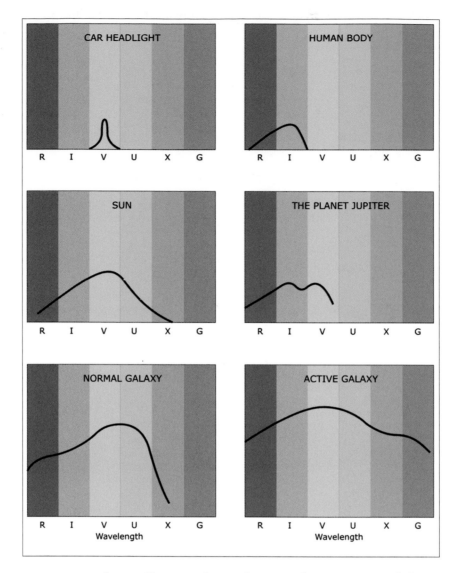

gamma rays, but still tiny—about the size of an atom—and they have one hundred to one hundred thousand times the energy of a visible photon. Ultraviolet rays are larger than x-rays—about the size of a molecule—and have up to one hundred times more energy than visible light. Visible photons have wavelengths that are larger than molecules, but smaller than infrared photons, which can be as tiny as a single cell or as large as the head of a pin. Infrared light carries up to one hundred times less energy than visible light. Microwaves are even larger—ranging from the size of an insect to the size of your foot—and they carry one hundred to one million times less energy than visible light. Radio waves are the biggest of all. They can be as tall as a child or as long as a soccer field, and they carry a million to a billion times less energy than visible light.

All objects that produce light spread their energy across the electromagnetic spectrum in a specific way. Figure 2.1 contains a series of graphs that represent the approximate brightness of energy across the electromagnetic spectrum for six objects. The first represents a light bulb, similar to a car's headlight, which produces visible light. The second represents the human body, which radiates mostly heat (infrared light). The third represents the surface of the Sun, which shines most brightly in the green-yellow portion of the visible spectrum but also produces ultraviolet, radio, and infrared light. The Sun's hot corona, which can be seen during a solar eclipse, has a temperature of several million degrees Celsius and produces x-rays. The fourth graph represents the planet Jupiter, which is cooler than the Sun and gives off mostly infrared and radio light. The fifth graph represents a normal galaxy, which emits radio, infrared, visible, ultraviolet, and a small amount of x-ray light. The sixth represents an active galaxy that shines brightly at all wavelengths.

As we can see from the active galaxy represented in figure 2.1, astronomers need information from the entire electromagnetic spectrum. Figure 2.2 illustrates this in a different way with a series of cartoons. The first six panels in this series represent all of the light that a single astronomical object produces in each part of the electromagnetic spectrum. The colored shapes represent the different parts of the spectrum—red shapes are radio photons, orange shapes are infrared photons, yellow shapes are visible photons, green shapes are ultraviolet photons, blue shapes are x-ray photons, and violet shapes are gamma-ray photons. For each set of shapes (each type of light), the pattern is uncertain. Observers who can see only one of the panels, or even two or three panels, can only infer the true nature of what they are looking at. Yet when they are added together in the seventh and final panel, one sees all of the shapes (photons of light) and a recognizable pattern emerges.

To truly understand the violent and exotic objects in the universe, we need to search for x-rays. Celestial objects are made up of normal matter, the stuff of everyday life, and matter is made of atoms. Atoms, in turn, are made of particles called electrons, neutrons, and protons. X-rays are produced when atoms are heated to extreme temperatures, exposed to intense magnetic fields, or subjected to immense pressures.

When temperatures soar to a million degrees, electrons have so much energy that they are constantly on the move. In releasing some of this energy, they generate x-ray photons. Electrons that suddenly smack into a dense material will release x-rays. Electrons that are moving in a strong magnetic field can be pulled in different directions, which causes them to spiral rapidly and emit x-rays. Electrons can also force light to change its energy. If a radio photon runs into a fast electron, the electron will "kick" the photon, giving it more energy and turning it into an x-ray photon. Electrons that are imprisoned within atoms can also emit x-rays when they jump from one orbit to another.

X-rays can be made by anything from a normal star to a black hole. X-rays originate from the boiling surface of the Sun, and they arise from the sizzling surface of comets as they fly into the solar wind. X-rays are expelled by stars in their death throes and afterward, as shocks from the massive explosion propagate through the gas surrounding the star. X-rays are produced when compact and massive stars devour the atmospheres of their larger companions in binary star systems. Similarly, x-rays arise from dead stars (black holes and neutron stars) if material is falling toward them. X-ray fireworks are created when stars are born and die in rapid succession as a result of galaxies colliding with each other. X-rays are emitted by gas heated to millions of degrees as galaxies plow past each other in their gravitational waltz. Finally, x-rays are produced deep within the hearts of galaxies near exotic, massive black holes, which eject powerful jets in a blaze of energy.

So there are x-rays flying all around the universe. How do we see them? They are best captured with a detector that counts each individual x-ray photon, as opposed to using an antenna, which is how radio waves are detected. An x-ray picture is not the same as a medical x-ray. The doctor's image is made by using a small, controlled source of x-rays to send them through your body, where they are absorbed by your bones. A medical x-ray is a negative image because what you see recorded on film are the black shadows where the x-rays were absorbed, or removed, from the original source of light. In contrast, astronomical objects emit x-rays just like the source of radiation that the doctor uses. An x-ray picture of an astronomical object is not a negative but a positive image, where the x-rays emitted by the object are captured by a type of detector that acts like a digital camera.

X-rays from the Sun were first detected in 1949 by American scientists from the U.S. Naval Research Laboratory, who launched Geiger counters—detectors designed to measure radioactivity—aboard a German V-2 rocket. Their successful experiment proved that x-ray astronomy was possible, at least for studying the Sun. But when astronomers began designing x-ray experiments to detect other astronomical sources, they worried that these sources might be so faint that they would be difficult, if not impossible, to detect. This worry proved to be unfounded, for in 1962, a rocket carrying an x-ray telescope built by a team of scientists and engineers led by the Italian-born physicist Riccardo Giacconi discovered the second celestial x-ray source in the constellation of Scorpius (fig. 2.3). Giacconi won the Nobel Prize for physics in 2002 for his pioneering work in x-ray astronomy. This discovery proved beyond the shadow of a doubt that x-ray astronomy would be a powerful tool for understanding the universe.

The x-ray pictures in this book were taken by satellites flying high above Earth's atmosphere. These satellites contain telescopes that focus x-rays into an image in a way that is very different from the way in which an optical telescope works.

Optical telescopes use slightly curved glass mirrors to reflect visible light toward a single point called the telescope focus. X-rays have so much energy that they are absorbed by such mirrors. An x-ray photon can only be reflected if it hits a surface at a very small angle. So x-ray telescopes contain mirrors with their surfaces held parallel to the incoming x-rays. When x-rays hit, or graze, this surface, they are bounced toward the telescope focus, creating an image.

The first high-quality x-ray images were obtained in the 1970s. Two of the most important precursors to today's x-ray telescopes were those on board NASA's Skylab space station and Einstein Observatory. Skylab was launched in 1973 and carried eight coordinated telescopes to study the Sun's spectrum in x-ray (fig. 2.4a), ultraviolet, and visible light. Einstein, launched in 1978, was the first fully imaging x-ray telescope freely flown on a satellite. It provided an imaging quality similar to today's telescopes and obtained the first sensitive images of the x-ray sky (fig. 2.4b and fig. 2.4c). As a result of these spectacular early images, the Einstein mission completely revolutionized our view of the x-ray universe.

The most powerful x-ray telescope ever flown is called the Chandra X-ray Observatory (fig. 2.5a). Chandra was launched by NASA in July 1999, and is named after the late Indian American scientist Subrahmanyan Chandrasekhar, winner of the 1983 Nobel prize for physics. Chandra is the largest satellite ever launched by the space shuttle and it flies above the Earth 200 times higher than the Hubble Space Telescope. Two other contemporary x-ray telescopes with images in this book are the ROSAT, or Roentgen satellite (1990–99), named after the German discoverer of x-rays, Wilhelm Roentgen, and the XMM-Newton, or X-ray Multi-Mirror Mission (fig. 2.5b), named after the British physicist Sir Isaac Newton.

X-ray telescopes have brought us many beautiful astronomical images. But since we can't see x-ray light, these images are not printed in true colors. This means that we have to decide what color means when displaying an x-ray image. X-ray light doesn't possess specific colors, but each photon of x-ray light has a specific energy, and so it is possible to code the x-rays with different colors corresponding to their energy. Every astronomical source also produces different numbers of photons from different locations, and so we can use color to express the relative intensity of x-ray light across an image rather than expressing the energy of the x-rays. This book uses both of these ways to display x-ray images.

Showing the intensity of the x-ray light rather than its specific energy is done with false-color images. Figure 2.6 shows three false-color images of the center of a nearby galaxy called NGC 253. Each image is made up of photons from within a specific energy range, while the colors in each image merely represent the intensity, or number of photons at every location. The image in the first panel (a) is made up of x-rays having low energies, the image in the second panel (b) is made up of x-rays having medium energies, and the image in the third panel (c) is made up of x-rays having high energies. The different colors range from fainter regions (purple) to brighter regions (red, white). Throughout this book, images that use color to show intensity will be referred to as intensity-colored images.

Color can also be used to indicate energy rather than intensity. Figure 2.7 shows the same panels as figure 2.6, but this time with the colors representing energy (red, green, and blue from first to last) rather than intensity. In this case, the color red represents x-rays with the lowest energies and the color blue represents x-rays with the highest energies. When these three images are combined, they create a single image where color represents x-ray energy (fig. 2.8). Throughout this book, images that use color to show energy will be labeled as energy-colored images. Energy-colored images provide the most information because they can tell us the energy and the intensity of the x-ray light at the same time.

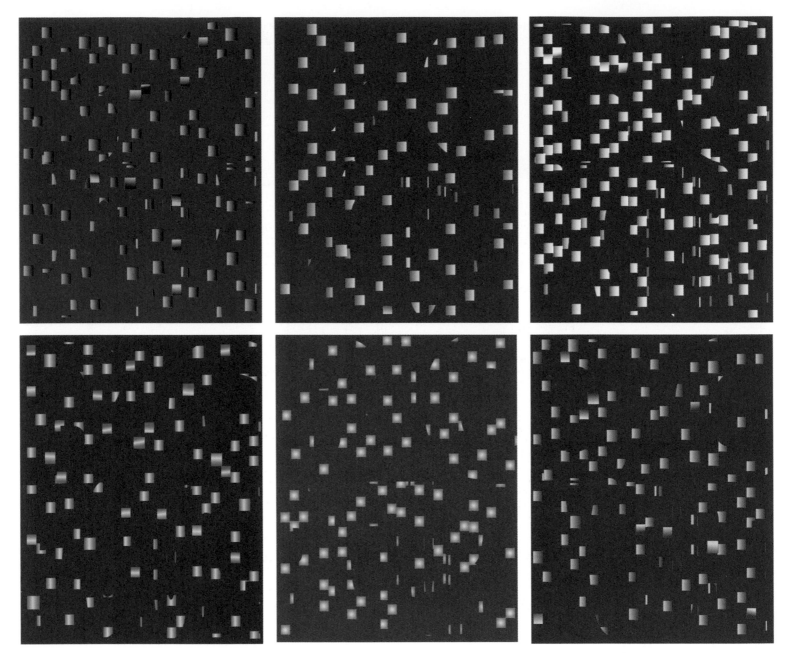

FIGURE 2.2. What our eyes are missing. This set of cartoons uses simple shapes and colors to illustrate how much of our universe we would be missing if we could only see it with our eyes. The shapes and colors are meant to represent the various types of light that make up the electromagnetic spectrum. Red shapes represent radio light, orange shapes represent infrared light, yellow shapes represent visible light, green shapes represent ultraviolet light, blue shapes represent x-ray light, and violet shapes represent gamma-ray light. Notice also that each set of shapes contains a subtle change of color. This is meant to represent the fact that even within a given region of the electromagnetic spectrum, light can be divided into still smaller pieces, the way visible light can be divided into the colors of the rainbow. The final drawing shows what you get when you put it all together (opposite).

FIGURE 2.3. Scorpius X-1. The first x-ray source detected besides our Sun, Scorpius X-1, was discovered in 1962 with an x-ray detector flown on a rocket. It was given this name because it was the brightest source of x-rays in the constellation Scorpius. First thought to be a single x-ray star, Scorpius X-1 is now known to be a system of two stars producing x-rays as one star devours the outer layers of the other star. The open circles in the top line of the graph (a) show the increase in the brightness of x-ray light at the location where Scorpius X-1 was discovered. Bright sources of x-rays can be useful for other experiments. The image taken with the Roentgen Satellite (b) shows the shadow of the Moon passing in front of Scorpius X-1. This observation proves that Scorpius X-1 is located outside our solar system.

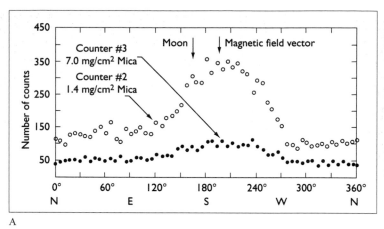

A

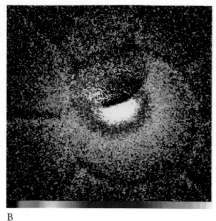

B

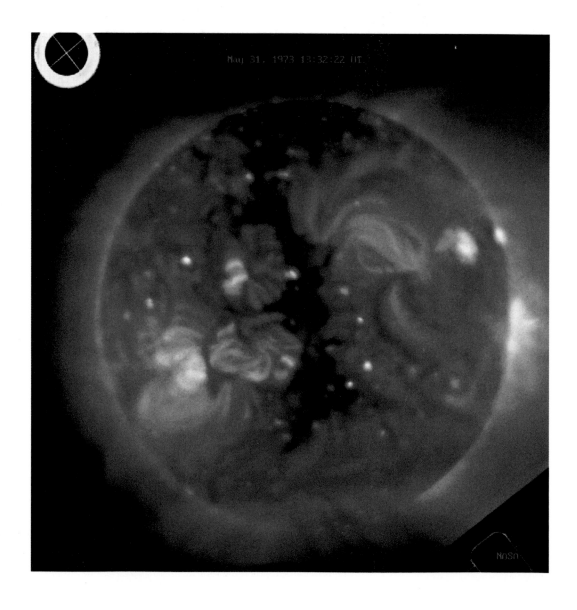

FIGURE 2.4. Early x-ray images. Obtained in the 1970s, these images are from x-ray telescopes that are the precursors to today's telescopes. The picture of the x-ray Sun (a) was taken with a telescope that flew on Skylab (1973–79). The long exposure of the x-ray sky (b) and the image of the supernova remnant Cassiopeia A (c) were obtained with the Einstein Observatory (1978–81). Compare these images to others later in this book to see the advances made with modern x-ray telescopes.

B

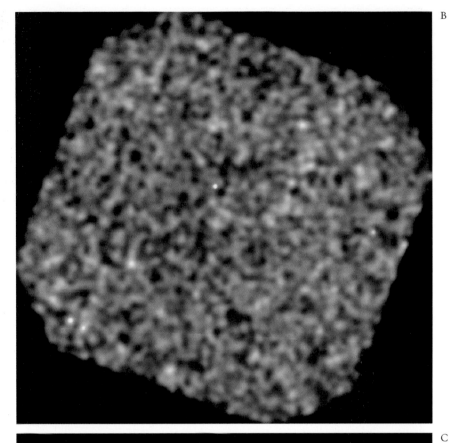

C

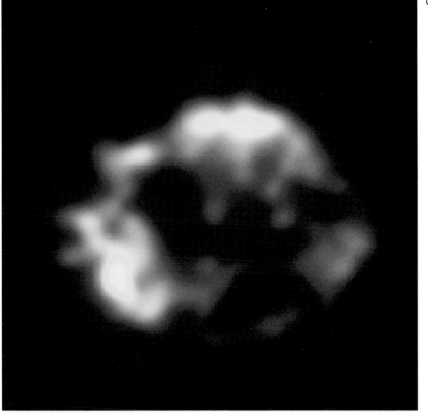

◄ A

◄A, B▲

FIGURE 2.5. Satellites. These images show artists' drawings of the Chandra X-ray Observatory in flight (a) and the XMM-Newton Observatory in flight (b). The two satellites were launched in 1999, and both are still in operation.

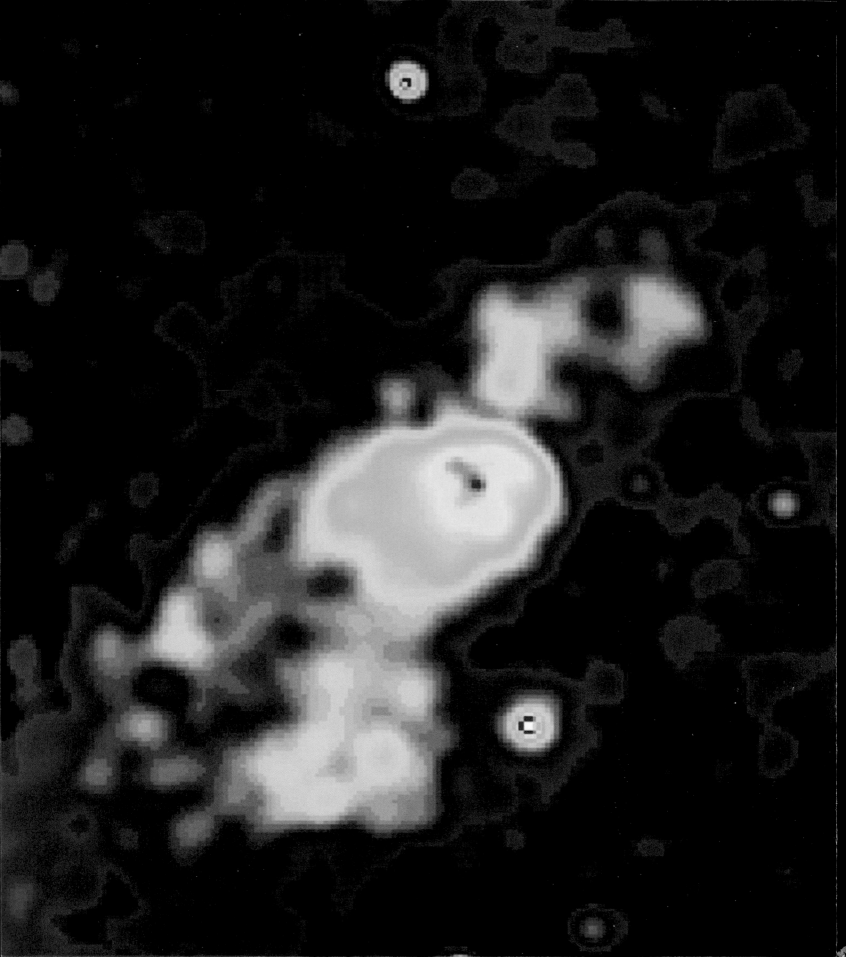

FIGURE 2.6. Intensity-colored images. These images are false-color images of a galaxy called NGC 253 split into three different bands of the x-ray spectrum. First is the image of low-energy, or soft, x-rays (a). Second is the image of medium-energy x-rays (b) and third is the image of high-energy, or hard, x-rays (c). The colors in these images have nothing to do with the energy of the x-rays, they are merely chosen to highlight the brightness of the light, with white, red, and yellow being the brightest areas and green, blue, and purple being the faintest areas. Images like these will be called intensity-colored images throughout this book. Intensity-colored images can have any color scheme.

B

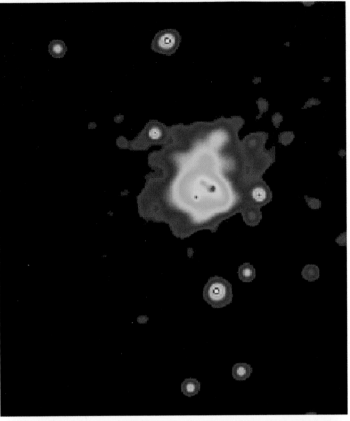

C

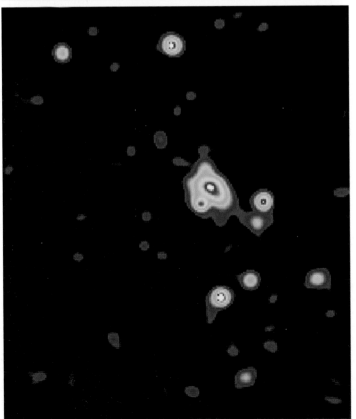

◄ A

What Our Eyes Can't See

Pages 38–40

FIGURE 2.7. Energy-colored images. These images are also false-color images of NGC 253 split into the same three energy bands in figure 2.6. However, this time the colors are chosen to correspond with energy. Red represents low-energy x-rays (a), green represents medium-energy x-rays (b), and blue represents high-energy x-rays (c).

A

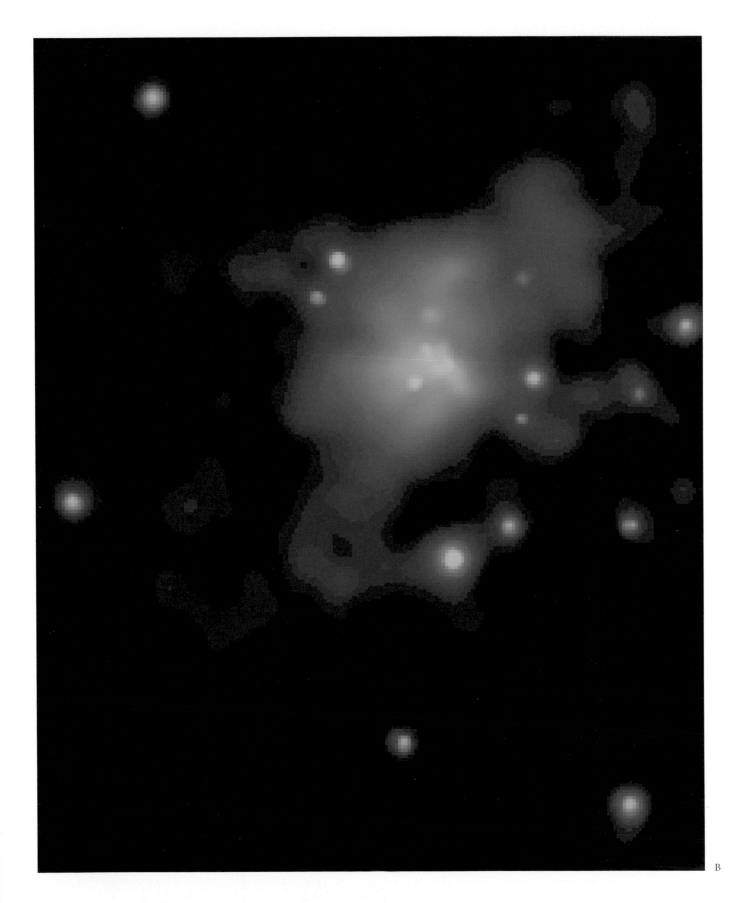

B

39

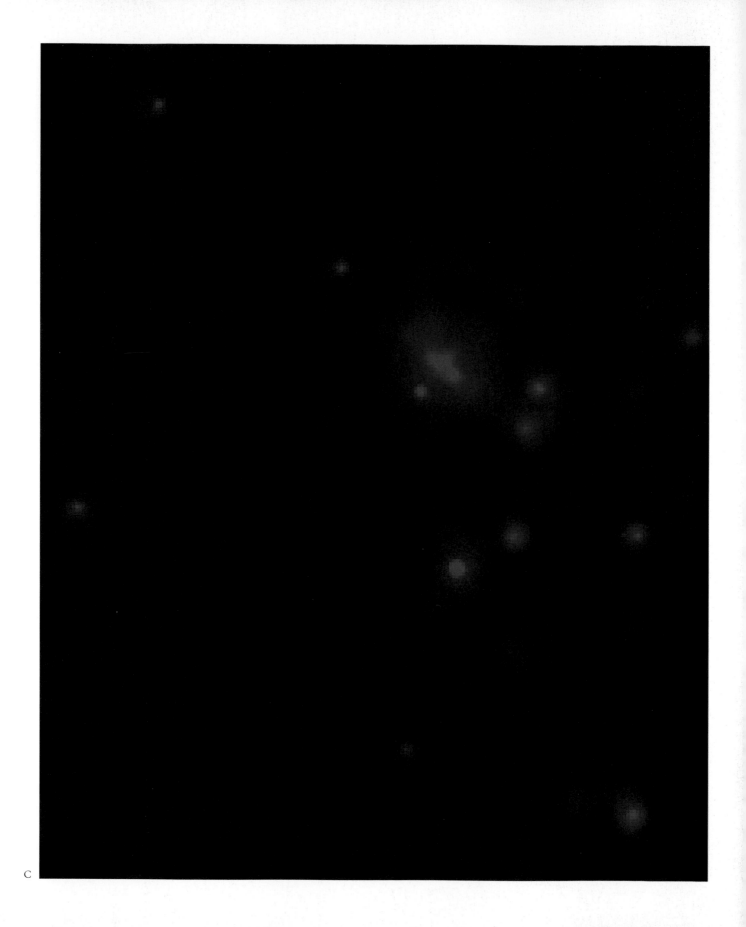

C

40

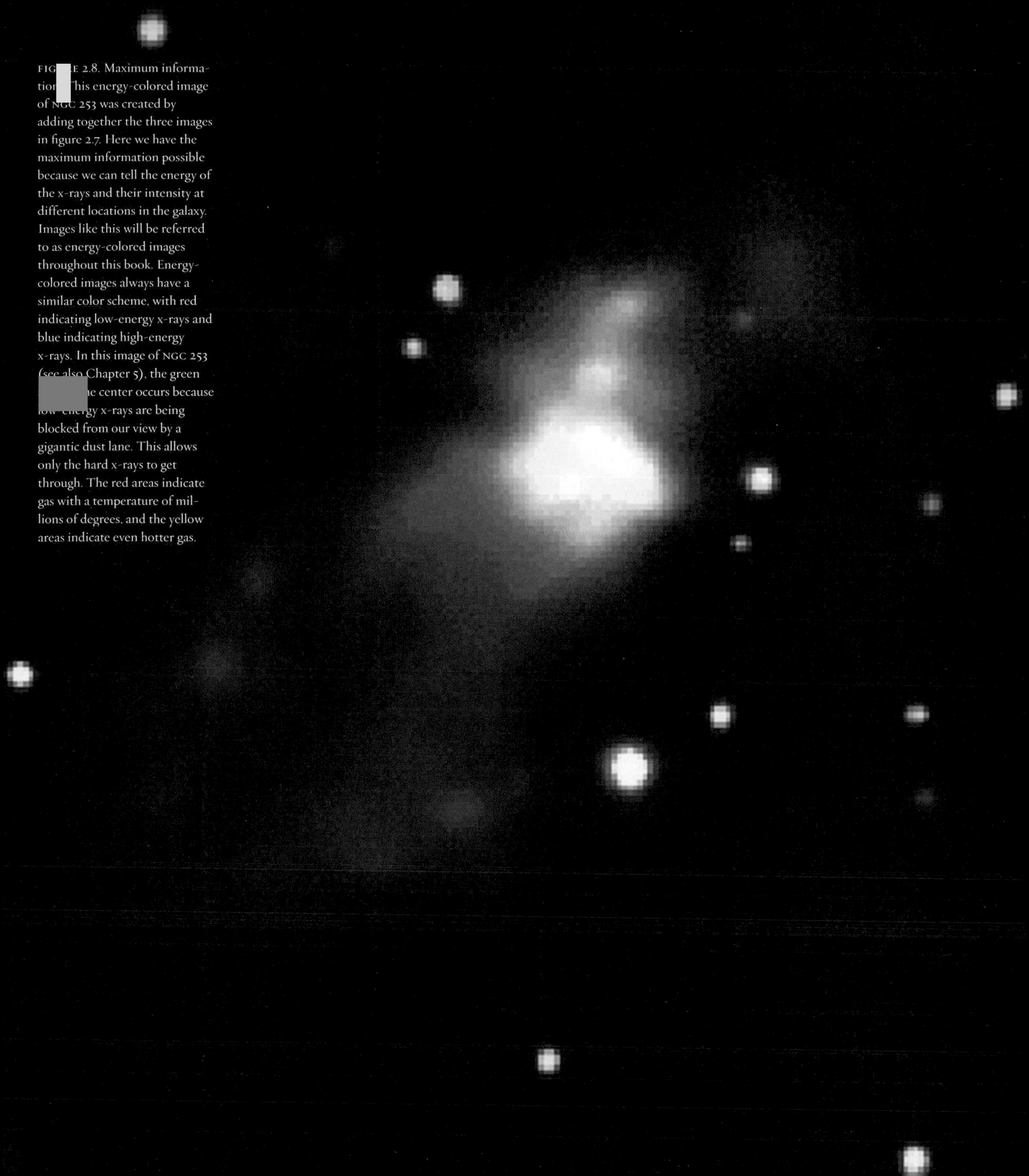

FIGURE 2.8. Maximum information. This energy-colored image
of NGC 253 was created by
adding together the three images
in figure 2.7. Here we have the
maximum information possible
because we can tell the energy of
the x-rays and their intensity at
different locations in the galaxy.
Images like this will be referred
to as energy-colored images
throughout this book. Energy-
colored images always have a
similar color scheme, with red
indicating low-energy x-rays and
blue indicating high-energy
x-rays. In this image of NGC 253
(see also Chapter 5), the green
at the center occurs because
low-energy x-rays are being
blocked from our view by a
gigantic dust lane. This allows
only the hard x-rays to get
through. The red areas indicate
gas with a temperature of millions of degrees, and the yellow
areas indicate even hotter gas.

3

From Cradle to Grave

OUR SUN is a vibrant source of x-rays, but that doesn't make it unique. The Sun is just like any other normal star, and stars give off x-rays throughout their lives. This means that astronomers can use their x-ray telescopes to trace the entire lifespan of stars, from birth to adolescence, through adulthood to old age, and finally, death.

The extremely hot outer layer of our Sun, called the corona, reaches temperatures of millions of degrees and is constantly alive with solar flares, loops, and prominences (fig. 3.1a). Solar flares are gigantic explosions on the surface of the Sun, and they can have as much power as a billion hydrogen bombs. Large flares can generate enormous bursts of x-rays that escape from the solar neighborhood and travel far into the solar system. If it weren't for the Earth's protective atmosphere, x-rays from the Sun would pose serious trouble; they could kill much of the life on our planet. In contrast to the x-ray view of the Sun, a visible image (fig. 3.1b) shows the Sun as we are used to seeing it—with its smooth surface, or photosphere, which is 1,000 times cooler than the corona. The ultimate source of the Sun's and any star's violent energy is nuclear fusion, a process by which the extreme temperatures and densities within the star's core combine hydrogen atoms into helium atoms. The Sun produces so much energy that one second of its output would provide a country like the United States with enough energy for the next several million years.

Stars are not born alone; they form surrounded by many siblings in giant stellar nurseries made up of clouds of gas and dust. These giant clouds contain enough material to make thousands to hundreds of thousands of stars. The process begins when a small portion of the parent gas cloud becomes dense enough to collapse under the weight of its own gravity. The cause of this cloud collapse might be shock waves from the explosion of a nearby star, a collision between the parent cloud and other clouds, or the effects of strong winds of radiation from newly born stars nearby. As the parent cloud collapses and breaks apart, the tiny cloud fragments make proto-stars, which continue to grow as long as they can gather more material from their surroundings. Once nuclear fusion ignites in their cores, these stellar infants can have birth weights ranging anywhere from a fraction to one hundred times that of our Sun.

A beautiful example of a stellar nursery is shown in figure 3.2. Here, the intense pressure of ultraviolet light from nearby hot and massive young stars has blown gas away and sculpted the parent gas clouds into giant, delicate gas pillars. These gigantic stellar incubators are several light years tall. If you look closely, you can see miniature balls of dense gas that had once been buried within the pillars at the tips of fingerlike features sticking out from the edges of the pillars. Many of these gas balls, which can be thought of as cocoons of gas, contain embryonic stars. The stars will stop growing when the cocoons are uncovered, and each star will emerge on its own once it heats the surrounding gas enough to blow away its cocoon.

Because the protective cocoon of gas and dust blocks visible light from these baby stars, it is not possible to witness their birth with optical telescopes. However, we can peer into stellar nurseries by searching for infrared, radio, and x-ray light, all of which can pass right through the dense gas. The Orion Nebula is the closest massive star-forming region to Earth. Figure 3.3a shows a visible image of the central region that is about 2.5 light years across. The x-ray image (fig. 3.3b) shows about a thousand young stars within a larger area that is about 10 light years across. The bright stars in the center are less than one million years old. Some were born already weighing up to 30 times more than our Sun. Because these stars are young, they are constantly changing and can alter their brightness over a period of just a few days.

Figures 3.4 and 3.5 are x-ray portraits of other impressive and bustling star forming regions. The group of stars called NGC 3603 (fig. 3.4) contains hundreds of young stars in a region of star formation 20,000 light years away. These stars were born about two million years ago. Besides producing x-rays in their coronae, some young stars have outer surface layers that are unstable. If enough pressure builds up, these babies can blow off their outer layers in the form of hot gas plumes or stellar winds. X-rays may also be produced from shocks that occur when fast winds ram into surrounding dense material. The large Rosette Nebula, which covers an area of the sky four times the area of the full Moon, is shown in figure 3.5. In the energy-colored x-ray image of the center of the nebula, the red stars are bare young stars that produce lots of low-energy x-rays, whereas the blue stars are still surrounded by cool gas that blocks their low-energy x-rays from our view. Star-forming nebulae are valuable laboratories for studying the processes by which our own Sun and solar system were formed some 4.6 billion years ago.

A star's lifetime depends on its mass. Stars like our Sun live for billions of years, but more massive stars will have much shorter lifetimes, because they burn out more rapidly. As a massive star nears the end of its life, the huge inward gravitational pull on the outer layers of the star is no longer balanced by the enormous pressure of radiation pushing outward; the star rapidly boils material off of its surface. One such unstable old star is Eta Carina (fig. 3.6), which is currently 5 million times more luminous than our Sun and could explode at any time. The immense gas bubbles that have been ejected by the star shine brightly in visible light, but the star itself can only be seen in the x-ray image. The yellow horseshoe-shaped ring in the x-ray image is made up of gas that has been heated by a large explosion on the star's surface that happened over a thousand years ago.

The death of a star with roughly the mass of the Sun is a pretty quiet affair. Just like Eta Carina, it may go through periods of being unstable and lose several outer layers of gas, but the star itself will eventually fade away in a calm fashion. Once such a star has used up its fuel, it simply cools down and becomes what astronomers call a white dwarf. Figure 3.7 shows the Cat's Eye Nebula, made up of multimillion-degree gas that has been expelled from its central star. The bright star in the x-ray image (fig. 3.7a) is expected to be compressed by its own gravity into a white dwarf in only a few million years. The composite image (fig. 3.7b) shows how the hot x-ray gas fills the region between the star and the cooler material that has been blown off, seen here in visible light.

When a star is sufficiently massive, its life comes to a dramatic and fiery end in one of the most cataclysmic events in the universe: a supernova explosion. Supernova explosions can shine as brightly as an entire galaxy and are triggered by the collapse of massive stars under the force of gravity, releasing enormous amounts of energy in the process. When such a star explodes, most of its mass is ejected at immense speeds, and the force of the explosion generates a blinding flash of radiation and shocks, like sonic booms. The matter thrown off by the explosion plows through the surrounding gas with tremendous energy, gathering up more material as it goes and creating an expanding shell of remarkably hot gas and high energy particles. This shell, called a supernova remnant, produces intense radio and x-ray light for thousands of years. Because heavy elements are formed inside stars during the process of nuclear fusion, supernova remnants are one of the primary ways in which the heavy elements necessary to form planets (and people) are released into the universe.

The images in figure 3.8 reveal the amazing detail and

incredible variety of the turbulent debris created by supernova explosions. At the top left (fig. 3.8a) is Tycho's supernova remnant, the remains of an explosion that was recorded in 1572 by the Danish astronomer Tycho Brahe. The debris, which looks like fingers of gas, has a temperature of about 10,000,000°C, while the blue circular arcs of x-rays along the outer rim are shock waves produced by gas expanding at a rate of about 20 million miles per hour (9,000 kilometers per second). At the bottom left (fig. 3.8b) is the remnant G292.0+1.8, which is about 1,600 years old, with a shell that is about 36 light years across, expanding at a rate of about 2.7 million miles per hour (1,200 kilometers per second). At the top right (fig. 3.8c) is the remnant N132D, located in the Large Magellanic Cloud, a nearby companion galaxy to the Milky Way that is about 160,000 light years from Earth. Its horseshoe shape is due to shock waves from the collision of ejected material with surrounding gas clouds. At the bottom right (fig. 3.8d) is a 300-year-old remnant called Cassiopeia A. Cassiopeia A has a temperature of about 50,000,000°C and contains a variety of elements such as silicon, calcium, and iron. Each of these elements is blown out in a different pattern. Images of Cassiopeia A in the x-ray light of these three elements are shown in figure 3.9. Notice that the images all look different, which tells us that the temperature and composition of the gas changes in different regions of the remnant.

Perhaps the best-known remnant of a star's explosion is the Crab Nebula. The explosion of the central star was seen from Earth and recorded by Chinese astronomers in AD 1054. The energy of this explosion was so stupendous and the star was so bright that it was visible during the daytime. The Crab Nebula is about 6,000 light years from Earth, is about ten light years across, and is expanding outward at approximately 3 million miles per hour (1,300 kilometers per second). Figure 3.10 shows the nebula in radio, infrared, visible, and x-ray light. The outer edge of the structure in the radio image represents the outer edge of the nebula. At the center of the nebula, and seen best in the x-ray image, is a rapidly spinning star that emits pulses of radiation 30 times per second. Spectacular ringlike structures surround the star and two jets blast away in opposite directions. The image in figure 3.11 shows the relationship between the x-ray, radio, and visible radiation and illustrates the complete picture that can only be obtained by putting information from all wavelengths together.

The central source that powers the x-ray emission in the Crab Nebula is a neutron star. Neutron stars are the remains of massive stars that have exploded and consist of what's left of the star's collapsed core. Spinning neutron stars are called pulsars. When a pulsar is located within a supernova remnant, the nebula becomes more than just a historical record of a star's death; it is actively powered by a source of high-energy particles. A Chandra image of the supernova remnant G54.1+0.3 (fig. 3.12) reveals a bright ring of high-energy particles around its central pulsar. The pulsar is spinning at a rapid seven times per second, and produces an enormous flow of energy, which creates two jetlike structures blasting away from its poles. High-energy particles stream outward from the jets into the extended nebula.

A very different pulsar, called the Vela pulsar, is shown in figure 3.13. The small image shows clouds of multimillion-degree gas that are part of a much larger remnant of the supernova explosion, which occurred 10,000 years ago. The entire remnant is expanding outward at a speed of about 250,000 miles per hour (110 kilometers per second). The Vela pulsar, viewed close up in the larger image, is surrounded by a tornado-like structure of bright rings and jets. The remarkable thing about the Vela and Crab pulsars is that such highly ordered disklike and jetlike structures can exist within the chaotic environment created by a supernova explosion. Even though the original star has died, the object that remains can still exert a powerful influence on its environment.

Star deaths that are even more extreme result in the ultimate ending, a black hole. Black holes represent the greatest concentrations of matter known in the universe and they are responsible for all sorts of bizarre behavior that we can see with our x-ray telescopes, which is discussed throughout the rest of this book. ✳

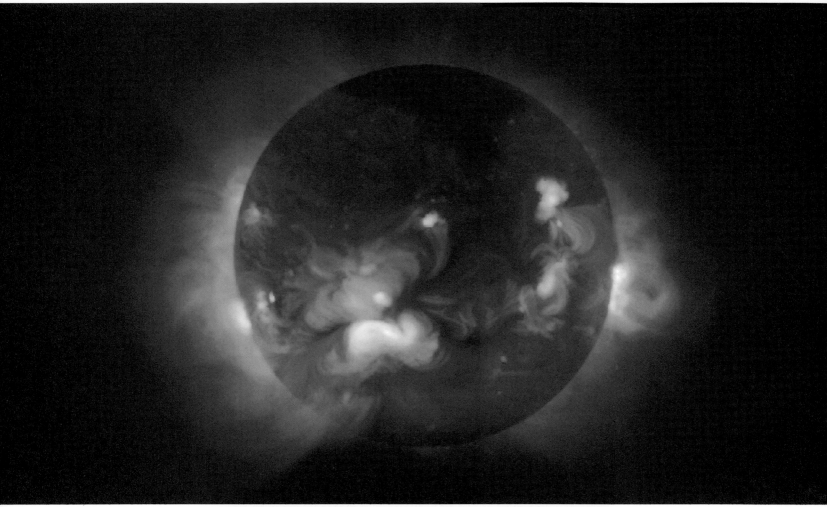

A

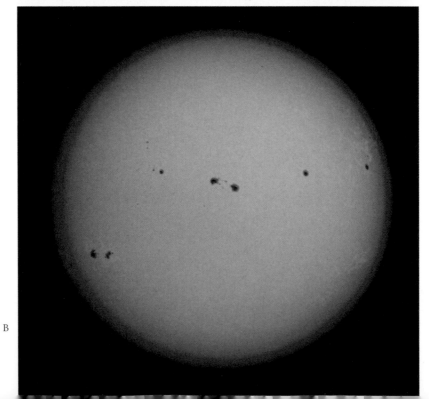

B

FIGURE 3.1. Our Sun. X-ray (a) and visible (b) pictures of the Sun, which contains 98 percent of all the mass in our solar system. The Sun is a giant ball of gas, so it doesn't have a well-defined surface. The region that we call the surface is where the Sun becomes so thick that light from beneath that region can't escape—this is the photosphere, which has a temperature of about 6,000°C. The outer hot corona, on the other hand, has a temperature of millions of degrees and emits x-rays. This region of the atmosphere launches loops and prominences and extends from the photosphere into the solar system.

FIGURE 3.2. A stellar nursery. This famous Hubble Space Telescope image of a stellar nursery called the Eagle Nebula shows newborn stars emerging from dense, compact cocoons of cool gas. This majestic star-forming region is 7,000 light years away from Earth and the giant towers of hydrogen gas are light years in length. Our solar system may have formed from a cloud very similar to these some 4.6 billion years ago.

FIGURE 3.3. Star babies. The Orion Nebula star cluster is located about 1,600 light years away in the constellation Orion the Hunter, seen overhead in the Northern Hemisphere during long winter nights. The giant Orion Nebula, first discovered in 1610, is the nearest massive region that is forming stars. The visible image (a) is about 2.5 light years across and shows a small portion of the entire giant gas cloud. The intensity-colored x-ray image (b) is about 10 light years across. The young bright stars in the center are just babies—less than a million years old. The dark vertical and horizontal lines in the image are part of the x-ray camera where light is not collected.

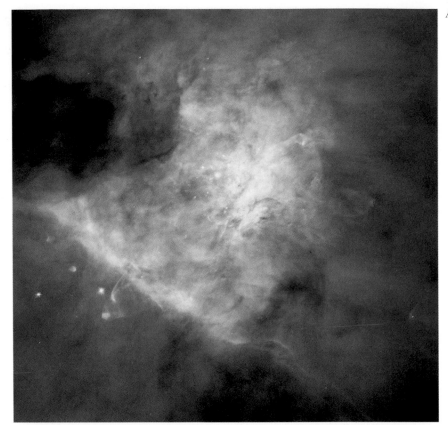

A

B

FIGURE 3.4. Star toddlers. The intensity-colored x-ray image of NGC 3603 shows another lively region of star birth, in this case about 20,000 light years from Earth. The reddish colors indicate the brightest stars, and green indicates the dimmest stars. The x-rays appear to come from the winds being blown off of massive young stars, some of which were born only 2 million years ago. The strong winds from these stars have cleared away most of the material around them.

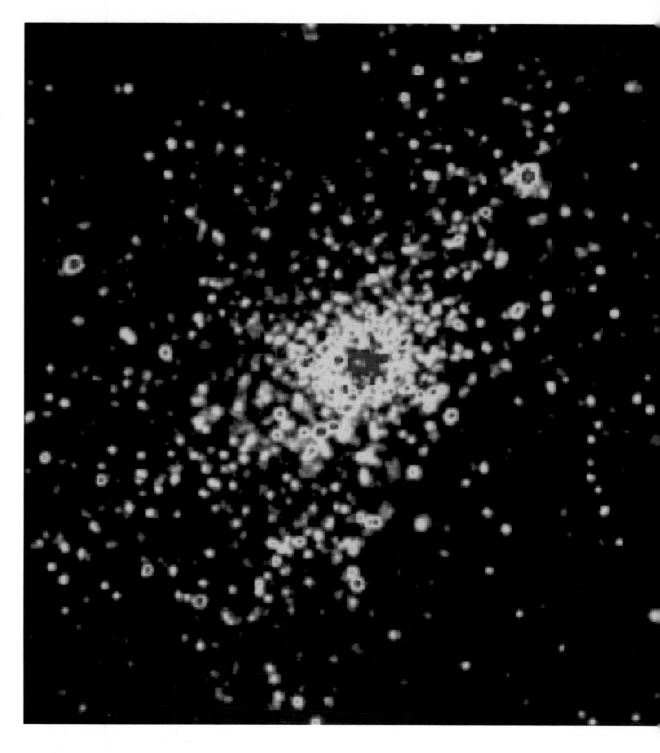

FIGURE 3.5. A nest of stars. The beautiful Rosette Nebula is 4,000 light years away and about 100 light years across. The visible image (a) shows a central hole where the strong winds from the young stars have cleared out most of the gas. The energy-colored x-ray image (b) shows a group of hot young stars in the center of the nebula that are about 4 million years old. The red stars have blown away their cool gas cocoons, and so we see all of the x-ray light that they produce. The blue stars are still surrounded by cool gas that absorbs low-energy x-rays, and so part of the x-ray light is blocked from our view. The faint red light that is spread throughout the region is hot gas from the stellar winds of the most powerful stars.

A

B ➤

FIGURE 3.6. An unstable star. The massive Eta Carina is an unstable star located roughly 7,500 light years from Earth in the constellation Carina. This is currently the most luminous star in our galaxy, giving off energy at a rate 5 million times the rate of our Sun. This star is rapidly nearing the end of its life and is quickly boiling material off its surface. The visible image (a) shows two huge clouds of gas that have been blown from the star in an explosion 150 years ago and are moving outward at a speed of 1.5 million miles per hour (670 kilometers per second). These giant clouds are about as large as our entire solar system. The energy-colored x-ray image (b) shows the central star and a ring of hot gas that may indicate another explosion over 1,000 years ago.

A

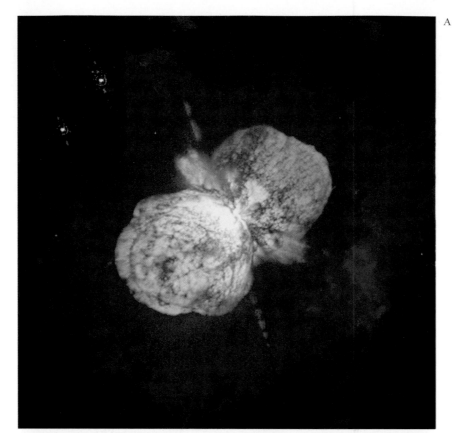

B

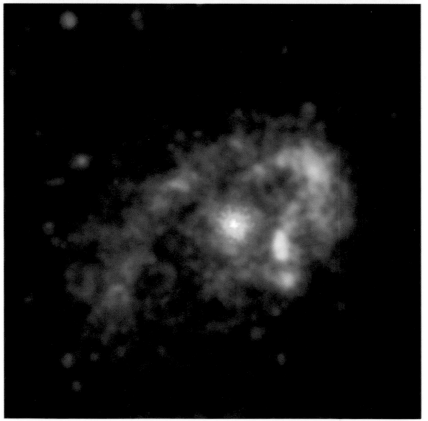

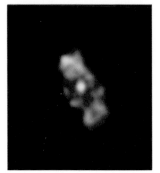

A

FIGURE 3.7. Cat's eye. The Cat's Eye Nebula is another dying star that is throwing off shells of glowing gas. This star, which is about the size of our Sun, will eventually collapse into a white dwarf. The intensity-colored x-ray image (a) shows that the star is surrounded by a cloud of hot gas. The composite image (b) allows us to see that the hot gas fills the inside of the visible nebula that surrounds the star. The visible nebula is made up of much cooler gas. This nebula is about 3,000 light years away and is estimated to be 1,000 years old. Its shape and structure tell us what is going on in the late stages of the star's life. Some astronomers have suggested that the two ringlike structures indicate a double, or binary, star system, with both stars throwing off their outer gaseous layers.

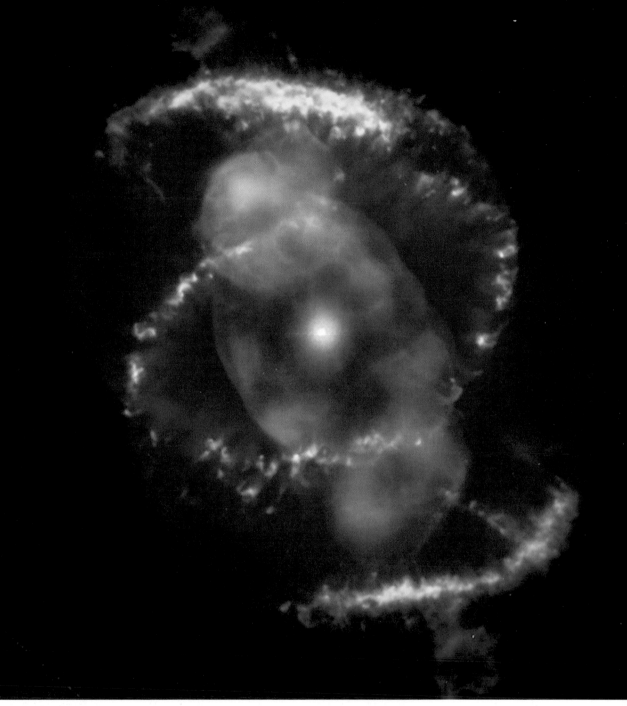

B

A

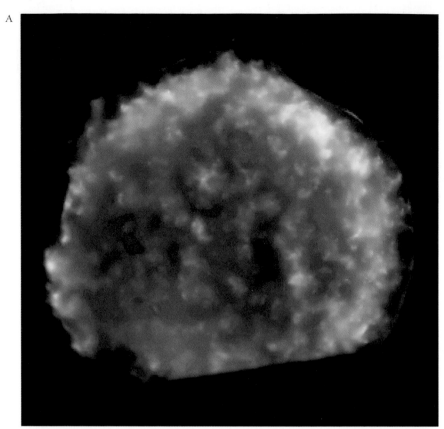

B

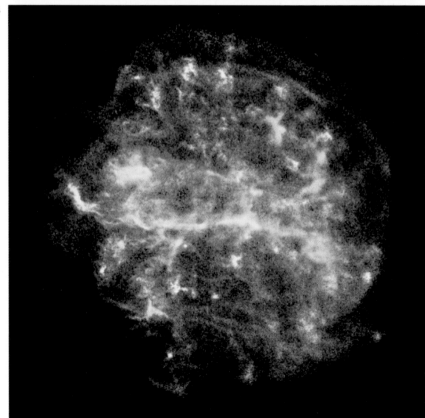

FIGURE 3.8. Stellar remains. This gallery of incredible x-ray images of supernova remnants illustrates some of the most dramatic events in the universe. All images are energy-colored. Tycho's supernova remnant (a) in the constellation Cassiopeia is 7,500 light years away. The nebula is 20 light years across, with an interior temperature of 10,000,000°C, rising to 20,000,000°C at its edges. This remnant has a lumpy look to it, as opposed to the knotty appearance of Cassiopeia A (d), but its outer edge is much smoother than those of the others shown here. The bottom of the Chandra image is cut off because the nebula is so large that it falls outside the region of the x-ray camera. The remnant G292.0+1.8 (b) is about 1,600 years old, and its shell of gas is 36 light years across. N132D (c) is 160,000 light years away, in one of the Milky Way's companion galaxies in the direction of the constellation Dorado. This remnant is expanding at a velocity of 3.7 million miles per hour (1,650 kilometers per second); it is about 3,200 years old, and its shell is about 80 light years across. The region to the upper left is a less dense region of space than the lower right, which explains why the nebula is expanding more rapidly in that direction. Cassiopeia A (d), a supernova in the constellation Cassiopeia, is 10,000 light years away and resulted from a star that exploded a mere 300 years ago. Expanding at a rate of about 4.5 million miles per hour (2,000 kilometers per second), the remnant is about 10 light years across, with a temperature of 50,000,000°C.

C

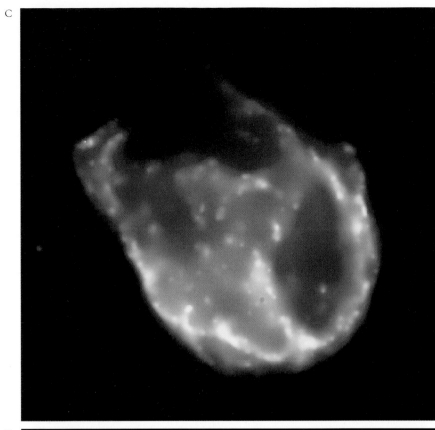

D

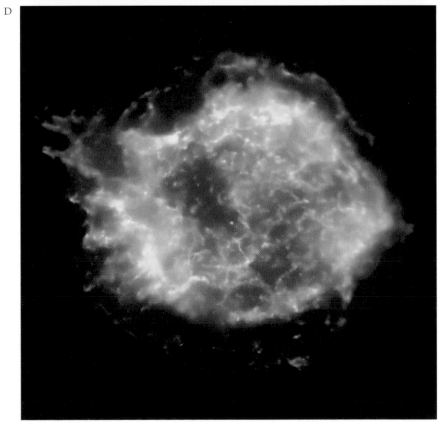

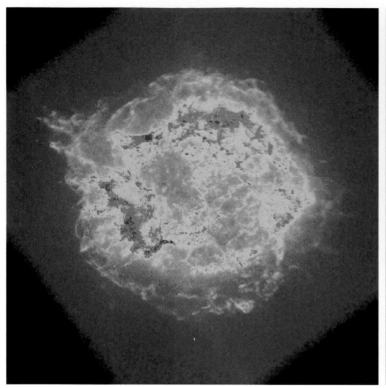

A

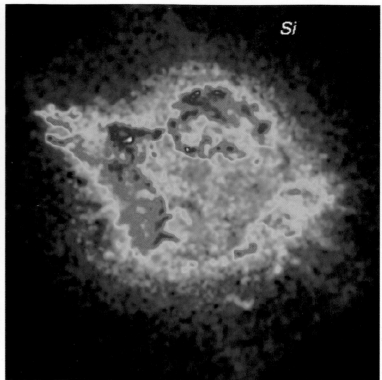

Si

B

FIGURE 3.9. Heavy elements. Most of the elements available to form stars and planets in the universe were produced in supernova explosions. The intensity-colored Chandra images of Cassiopeia A show some of the heavy elements ejected in this supernova explosion. The full x-ray image (a) is shown here in intensity for comparison with the others. X-rays from silicon (b) are spread throughout the nebula fairly evenly, but there is also a jet breaking out of the shell to the upper left, which suggests that the explosion of the star may not have been even in all directions. X-rays from calcium (c) are much clumpier than silicon, while iron (d) is located in even fewer places. There is also a large region to the left that is mostly iron. These images help astronomers know where the elements came from within the star just before the explosion, how the explosion may have occurred, and how the elements are now being spread out in space.

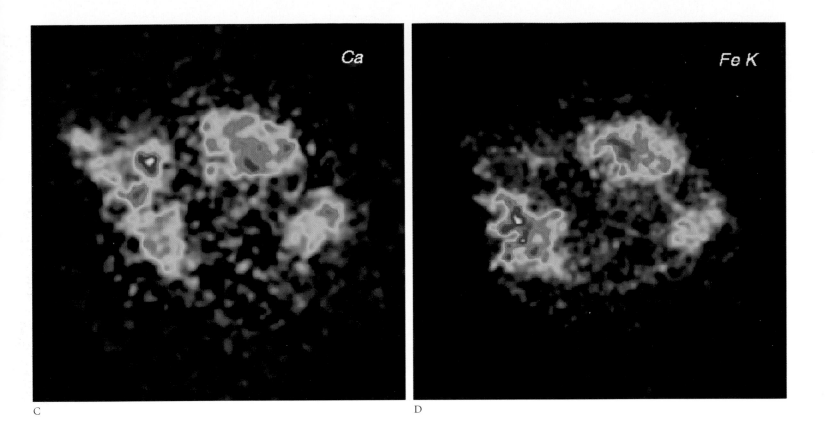

C

D

FIGURE 3.10. The Crab Nebula.
The Crab Nebula in the constellation Taurus, which is well
known to astronomers, is 6,000
light years away. The nebula is
what remained after a brilliant
stellar explosion in AD 1054,
recorded by Chinese astronomers. During its brightest phase,
it was visible during the day and
was as bright as the full Moon at
night. These images show the
nebula at radio (a), infrared (b),
visible (c), and x-ray (d) wavelengths. The images are not all to
scale—the x-ray image is about
twice its real size here compared
to the others. The intensity-
colored x-ray image shows a
rapidly spinning neutron star at
its center (a pulsar), which emits
radiation 30 times each second
as it spins.

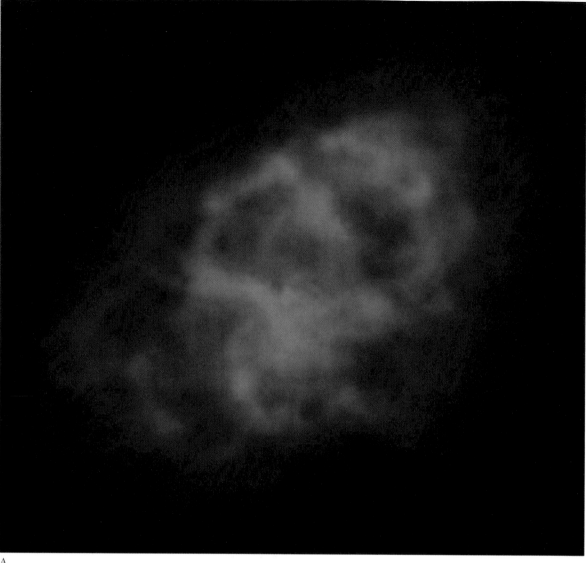

A

B ➤

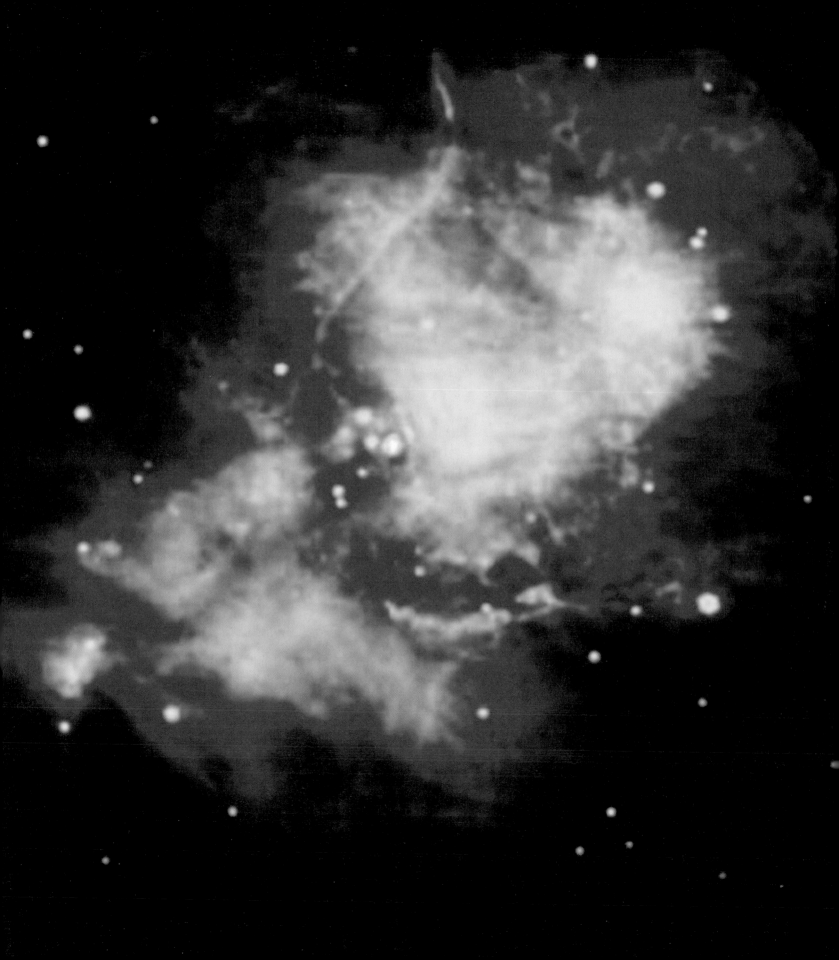

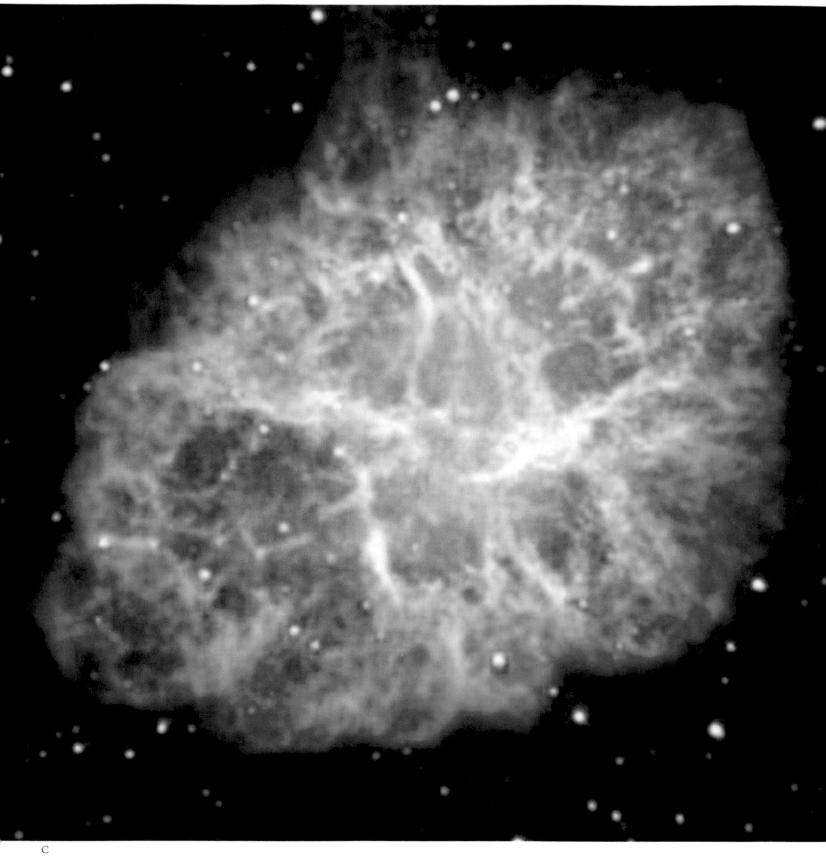

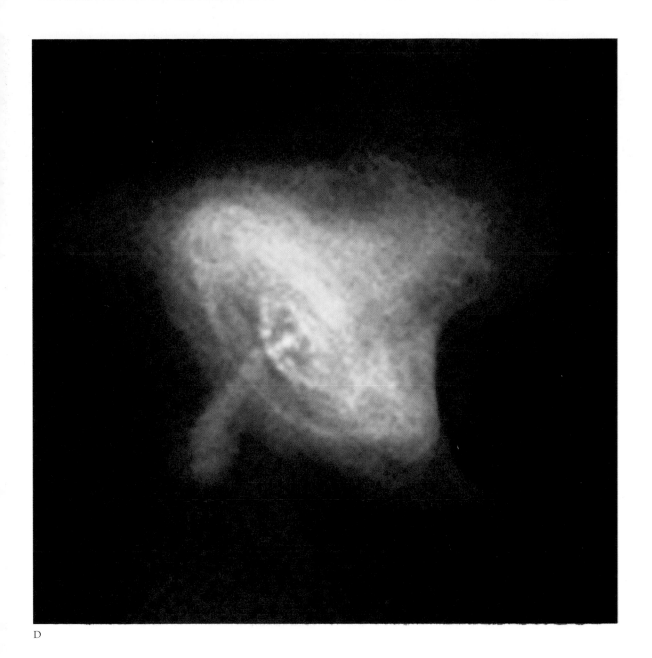

D

FIGURE 3.11. The Crab Nebula, full view. This composite visible, radio, and x-ray image of the Crab Nebula shows the images all to scale. The radio emission comes from the furthest out, while the visible light fills the nebula, and the x-ray disk and jets fit within the visible and radio nebula. Without the x-ray image, the spectacular behavior of the central pulsar, or rapidly spinning neutron star, would be very difficult to see.

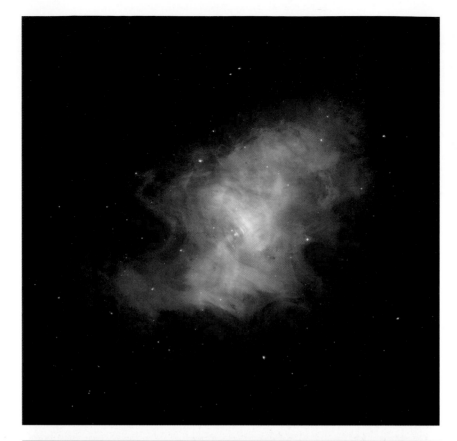

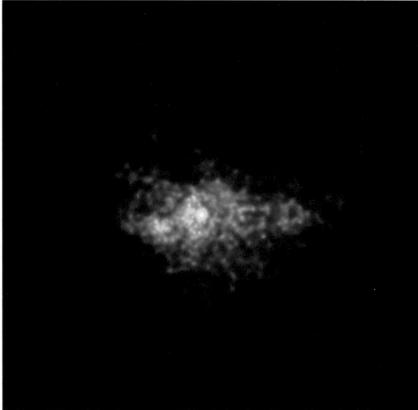

FIGURE 3.12. A pulsar nebula. This intensity-colored image of the supernova remnant SNR G54.1+0.3 shows a large nebula created by the central pulsar with a ring and two jetlike features, similar to the Crab Nebula. These features are made up of high-energy particles from the spinning neutron star, which rotates seven times each second. The electric field from the pulsar causes the particles to be accelerated and blasted away in the jets. This nebula is 16,000 light years away and about 6 light years across.

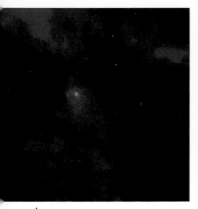

A

FIGURE 3.13. High-speed motion. The nebula surrounding the Vela pulsar looks very different from the Crab Nebula and SNR G54.1+0.3. The red cloudlike structure in the intensity-colored x-ray image (a) is hot gas left over from the giant supernova explosion that produced the pulsar about 10,000 years ago. The entire remnant is about 100 light years across, which is 15 times larger than the region shown here. The pulsar is the small yellow source in the center, which is seen up close in (b). There are two bright rings and jets, and the pulsar nebula is actually bent backward due to the motion of the pulsar toward the upper right, in the direction of the arrow, at a speed of about 200,000 miles per hour (90 kilometers per second).

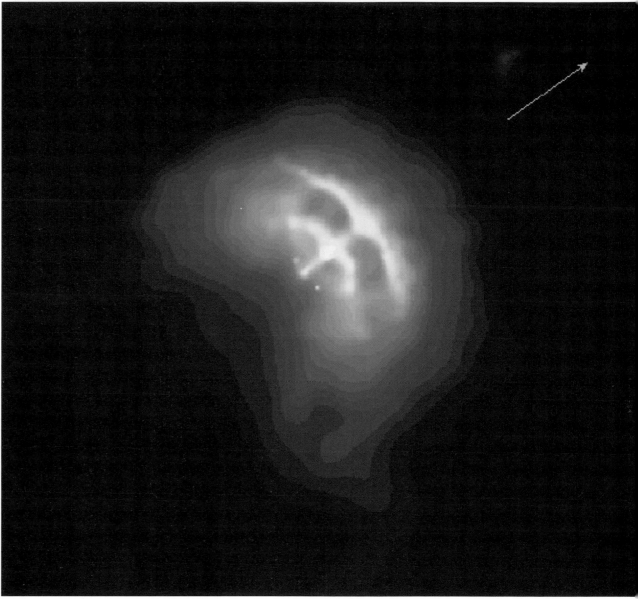

B

4

No Place Like Home

ARE GALAXIES mere collections of stars, gas and dust? Well, yes and no. It all depends on how you look at them. When astronomers examine a galaxy in visible light, they glimpse the locations of the normal stars and the lanes of dust that pervade the galaxy. But visible pictures miss some of the most fascinating parts of a galaxy—objects such as stars that have evolved past their normal lifetimes and now occupy regions where extreme physical conditions prevail. The most exotic of these "stellar endpoints" are black holes, whose gravity is so extreme that light cannot escape.

The term *black hole* describes a location in space where matter is packed so tightly that it produces an immense gravitational pull. The gravitational attraction of a black hole is so large that beyond a certain boundary, called the event horizon, nothing can escape. Because black holes don't give off light, we can't see them. That doesn't mean, however, that we can't find them. In the search for evidence of a black hole, one of the biggest clues is that a black hole's gravity distorts its surroundings and alters the path of light that passes by, as shown in figure 4.1. By searching for gravitational effects caused by a black hole, we can pinpoint its location.

Searching for x-ray light is the best way to look for black holes in galaxies, for many reasons. First of all, black holes and their gravity can wreak havoc on their environments by ripping entire stars apart. When this happens and the star's material falls into the black hole, x-rays are produced. Strong flares of x-ray light may be a sign that a black hole is feeding on something. Second, the large amount of energy carried by an x-ray allows it to pass through colossal amounts of dust and gas, which can act like curtains, blocking visible light. X-rays can escape the turbulent environments of black holes and then travel long distances through the thick disks of galaxies, so we have the chance to detect black holes no matter where they are.

The galaxy shown in figure 4.2 is the Andromeda galaxy and is located only about 2 million light years from the Milky Way galaxy, which is right inside our neighborhood by cosmic standards. The visible image shows the starlight from hundreds of billions of stars. However, this starlight is only a fraction of the total light produced by Andromeda. Figure 4.3 is an x-ray image of the central region of Andromeda. The pinpoints of light in this image are not normal stars but pairs of stars called x-ray binary systems. In an x-ray binary system, a compact remnant of a star (either a neutron star or a black hole) orbits a relatively normal star. The strong gravitational attraction of the neutron star or black hole sucks material from the surface of the normal star like a cosmic vacuum cleaner. This material then falls into orbit around the neutron star or black hole and heats up, giving off x-rays (fig. 4.4). The disk of material orbiting the neutron star or black hole is called an accretion disk.

Our Milky Way galaxy is very similar in shape and size to the Andromeda galaxy. Exotic objects such as black holes and x-ray binary systems exist here, just as they do in Andromeda, but it is a bit harder to see them directly. This is because our solar system is located right in the middle of the disk of our galaxy and we have to peer through this disk sideways to see neighboring objects, rather than looking down on top of the disk as we do in Andromeda. Looking through our galaxy's disk makes it difficult to see what's happening at its center when we search in visible light. The most obvious feature in a visible image of the Milky Way (fig. 4.5) is the large, dark dust lane that almost entirely blocks our view in some directions.

Luckily, x-rays can pass right through the Milky Way's dust lane. This fact is illustrated in the colorful Chandra x-ray image of the central region of the Milky Way shown in figure 4.6. This image provides a much clearer and more comprehensive picture of our galaxy's core than the visible image. The bright, pointlike objects are white dwarf stars, neutron stars, and black holes, mostly in binary systems, and these objects are surrounded by swirling clouds of hot gas at temperatures of many millions of degrees. The large, turbulent bubbles of hot gas were formed by winds boiling off the surface of young, massive stars and supernova explosions, and they contain the chemical elements that were released from dying stars. As the combined energy from these explosions builds over time, the gas is pushed outward and the chemical elements born inside stars are propelled throughout the galaxy.

The panels in figure 4.7 provide an even more complete view of the center of the Milky Way. Figure 4.7a shows the radio light, colored in orange; 4.7b shows the infrared light, colored in green; 4.7c shows the x-ray light, this time colored in blue. These three images are added together in figure 4.7d, which allows us to witness the intricate relationship between the objects that produce light at all of these different wavelengths. The hot x-ray gas coexists with cooler gas in the form of massive streamers and bubbles—material that is either exploding outward or being ripped apart by the violent forces at work in our galaxy's core.

A massive black hole may be lurking at the heart of the Milky Way. At the center of the galaxy is a compact radio source called Sagittarius A*, which is having a dramatic effect on orbiting stars. Astronomers have observed stars speeding rapidly around this area with velocities of up to 11 million miles per hour (5,000 kilometers per second). Not only that, but Sagittarius A* has never been seen to move from its location. It is not disturbed in the least by the stars that whip past, which suggests that it must be extremely massive in order to remain stationary under the violent conditions near the center of the galaxy. The rapid speeds of the stars indicate a large gravitational pull and suggest a remarkably massive object with a mass of 2.6 million Suns.

More evidence for a black hole at the center of our galaxy comes from x-ray observations. Figure 4.8 shows an enlargement of the central region of our galaxy. An immense cloud of hot gas surrounds Sagittarius A* (the white patch of light that is at the center of the image). When Chandra was watching it in 2000, the x-ray point source brightened in a flare that lasted only three hours, which is almost unfathomable for such a massive source. This quick change in x-ray brightness means that the source has to be small, sized no larger than the distance between the Earth and the Sun. The only object that could be so small and yet have a mass of 2.6 million times the mass of our Sun is a giant black hole. The flare was probably caused by the black hole ripping material away from a comet-sized object that fell into its vicinity. Figure 4.9 shows large streams of material that are continuously falling toward the center of the galaxy, providing a steady source of fuel for the black hole.

Besides the massive black hole at its center, evidence exists for smaller black holes in the Milky Way. As discussed earlier, small black holes tend to be located in x-ray binary systems, which can produce large opposing jets of high-energy particles streaming away from their accretion disks (fig. 4.10). Two of these systems are called SS 433 and XTE J1550. SS 433 is an x-ray binary in which the black hole's companion star has a mass about 20 times that of our Sun. Figure 4.11 shows that its jets end in two opposing high-speed lobes of hot gas, with a temperature of about 50,000,000°C. Material is ejected from near the black hole every few minutes in the form of blobs. These blobs travel outward like bullets at about one-quarter the speed of light. Once the blobs reach the end of the jet and start to slow down, they begin to collide with one another. These collisions heat the gas at the ends of the jets.

XTE J1550 is an x-ray binary system in which the companion star has a mass similar to that of our Sun. In 1998, XTE J1550 erupted in a strong x-ray flare, which was seen by a satellite called the Rossi X-ray Timing Explorer. Two years later, this source was examined with Chandra, which detected a western and then an eastern jet over a two-year period. The movement of the jets in a series of images from June 2000 to July 2002 (fig. 4.12) shows the evolution of the jets produced by highly energetic electrons that are catapulted from the black hole at half the speed of light. Scientists believe the jets were first formed when the flare occurred. In a period of only four years (1998–2002), the x-ray blobs moved a distance of three light years apart. By forming these intense jets, the black hole transfers its energy into its immediate environment.

While black holes at the center of our galaxy, in SS 433 and in XTE J1550, are some of the nearest black holes, they are by no means the most powerful examples. The Milky Way's central black hole is relatively quiet, appearing to flare up only occasionally, while XTE J1550 and SS 433 are examples of the smallest known type of black holes—stellar mass black holes. Much more powerful examples of black holes are certainly to be found, as we shall see later in Chapter 6. But we shall first examine some spectacular galaxies that produce much more x-ray light than the Milky Way.

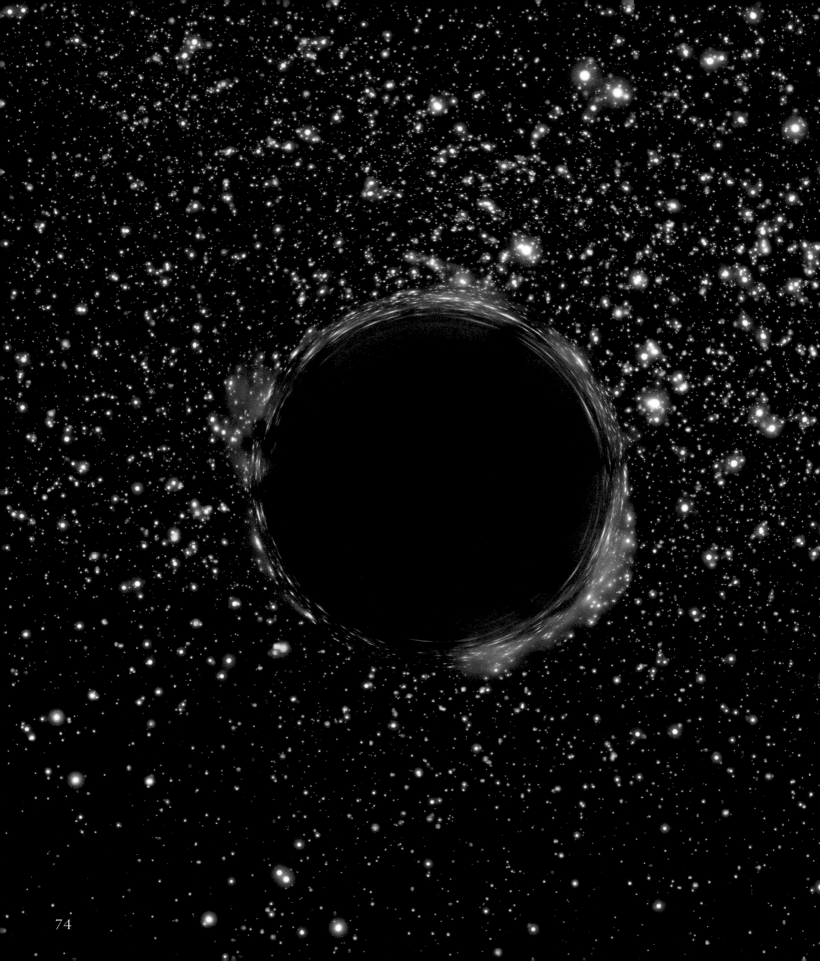

FIGURE 4.1 (opposite). A black hole. Black holes, which were first described mathematically as the result of the work of Albert Einstein in the early 1900s, were given their popular name by the American physicist John Archibald Wheeler in 1967. But the idea of a "dark star," the light from which is bent backward by its strong gravity, has been with us for more than 200 years. This artist's drawing shows what a black hole might look like as it passes between us and a distant star field. The hole is invisible because no light can escape, but its presence is obvious because it blocks the light behind it. The black hole also behaves like a lens, bending and stretching the starlight behind it into a ringlike outline surrounding its edge. It is this bending of light, along with the gravitational pull black holes exert on other objects, that allows astronomers to detect them.

FIGURE 4.2. The nearest large galaxy. The dazzling spiral galaxy known as Andromeda, or Messier 31, is the nearest large galaxy to the Milky Way and has been known since ancient times. It can be considered a sister galaxy to the Milky Way, since the two galaxies probably look very much alike. Andromeda, which contains roughly 300 bil-lion stars, is about 2 million light years away and 150,000 light years across (one and a half times the size of the Milky Way). You don't need a telescope to see Andromeda—it is easily visible to the naked eye. This collage shows visible images of the galaxy and the Moon placed next to each other to demonstrate their relative sizes in the night sky. The two small fuzzy objects are Andromeda's tiny companion galaxies.

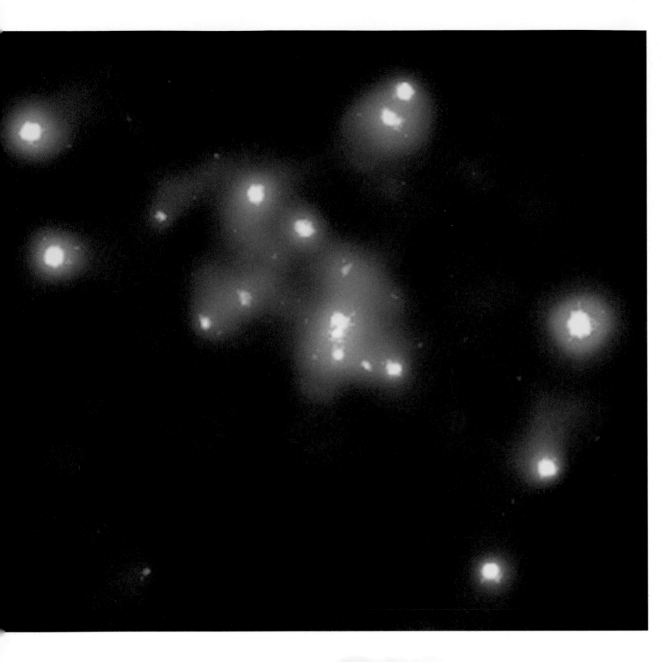

FIGURE 4.3. Bright lights. This intensity-colored x-ray image of the very center of the Andromeda galaxy shows a number of bright, pointlike sources of light. These are probably x-ray binary star systems, in which a normal star orbits either a neutron star or a black hole. A supermassive black hole weighing 30 million Suns is thought to lie at the position of the yellow dot just above the blue dot in the center of this picture.

FIGURE 4.4. A binary system. Illustration of a binary star system in which a normal star is orbiting a black hole. As matter is sucked from the star by the black hole's gravity, it forms an accretion disk around the black hole and is heated to temperatures of millions of degrees. This hot gas gives off x-rays.

Top:

FIGURE 4.5. The Milky Way, murky view. Our solar system lives within the plane of the Milky Way galaxy, about two-thirds of the way between the galaxy center and the its outer edge (roughly 25,000 light years from the center). It takes the Sun about 250 million years to make one orbit about the galaxy center. This panoramic visible image, which covers 90 degrees of the sky, makes the dramatic point that we have to look through an enormous amount of dust and gas to see along the plane of the Milky Way.

Bottom:

FIGURE 4.6. The Milky Way, unveiled view. The murky veil of dust is lifted in this energy-colored x-ray image of the inner region of the Milky Way galaxy, 400 by 900 light years in size. The galaxy is aglow with turbulent motion—x-rays from white dwarf stars, neutron stars, and black holes are bathed in a 10-million-degree gas created by massive stars boiling off their outer layers in stellar winds and shock waves from supernova explosions. This chaotic motion sends the elements created in stars throughout the rest of the galaxy.

A

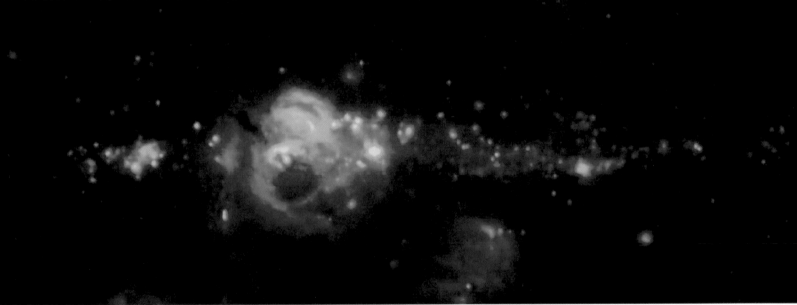

B

FIGURE 4.7. The Milky Way, multiple views. This series of images of the center of the Milky Way in radio (a), infrared (b), and x-ray (c) light shows how each electromagnetic window gives us a different view of what's going on. The composite energy-colored image (d) in radio (orange), infrared (green), and x-ray (blue) light highlights the intricate connection between these different sources of light. Here we see how the stars (x-ray light), gas (radio light), and dust (infrared light) are interrelated.

C

D

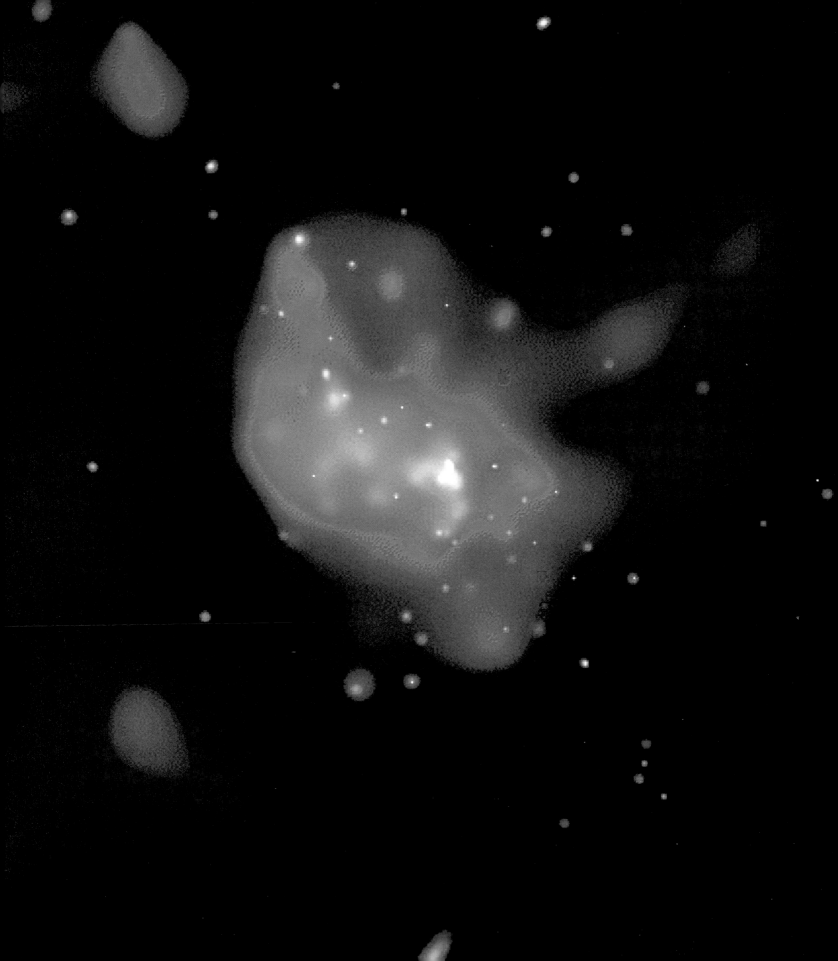

FIGURE 4.8 (opposite). A super-massive core. This intensity-colored Chandra image shows a region about 60 light years across at the heart of the Milky Way. Astronomers believe there is a supermassive black hole lurking here with a mass 2.6 million times that of our Sun. The bright white spot at the center of the image is the suspected location of the black hole. A quick flare of x-rays, which lasted about three hours, produced the spot. This flare indicates that material strayed close to the black hole and was probably devoured. To produce such a flare, the black hole would need to have swallowed something roughly the size of a comet.

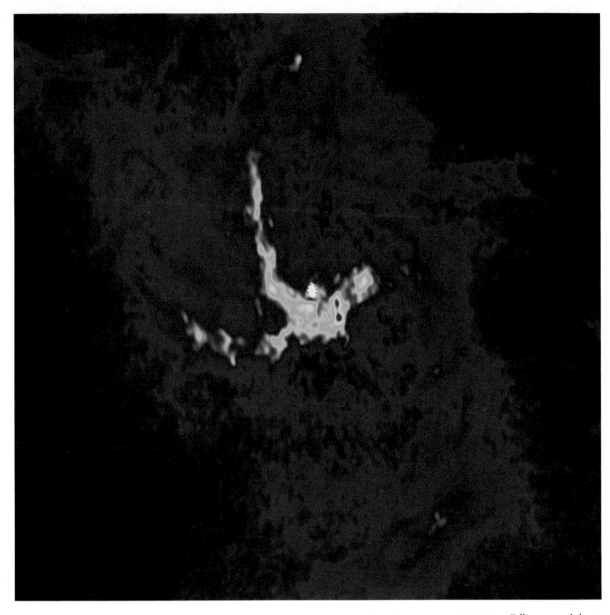

FIGURE 4.9. Falling toward the center of the galaxy. This intensity-colored radio image of the heart of the Milky Way shows long trailing arms of gas and dust streamers spiraling around the center of the galaxy. They contain material that is probably falling inward toward the galaxy's central black hole.

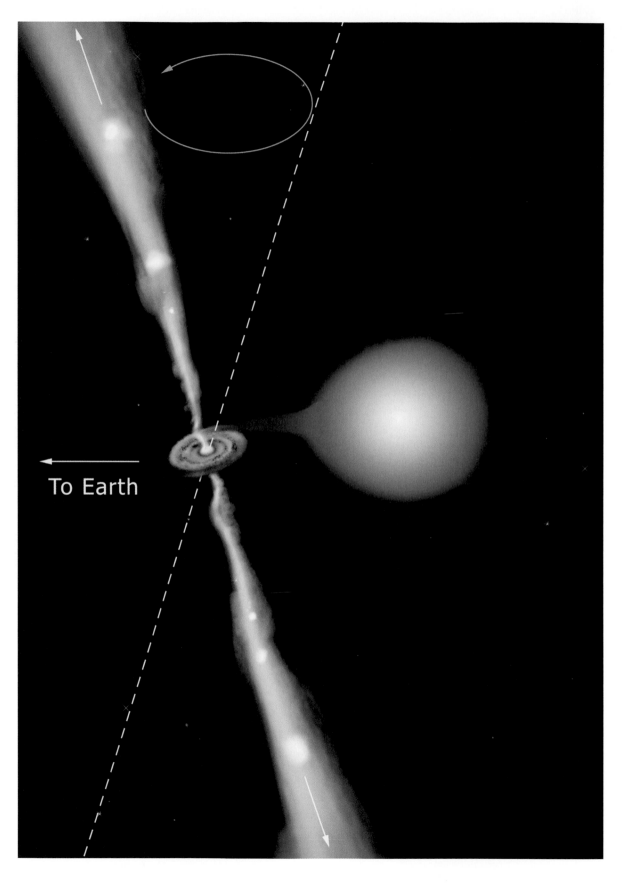

To Earth

FIGURE 4.10. Jets of high-energy particles. This artist's illustration shows a binary star system consisting of a massive star and a black hole. Material is pulled from the star and falls into a swirling accretion disk around the black hole. This material is then cooked to temperatures of millions of degrees. At the same time, the extreme forces caused by the black hole and the disk expel jets of high-energy particles in opposite directions. These jets are similar to, but much smaller than, the jets created by supermassive black holes in galaxies (Chapter 6).

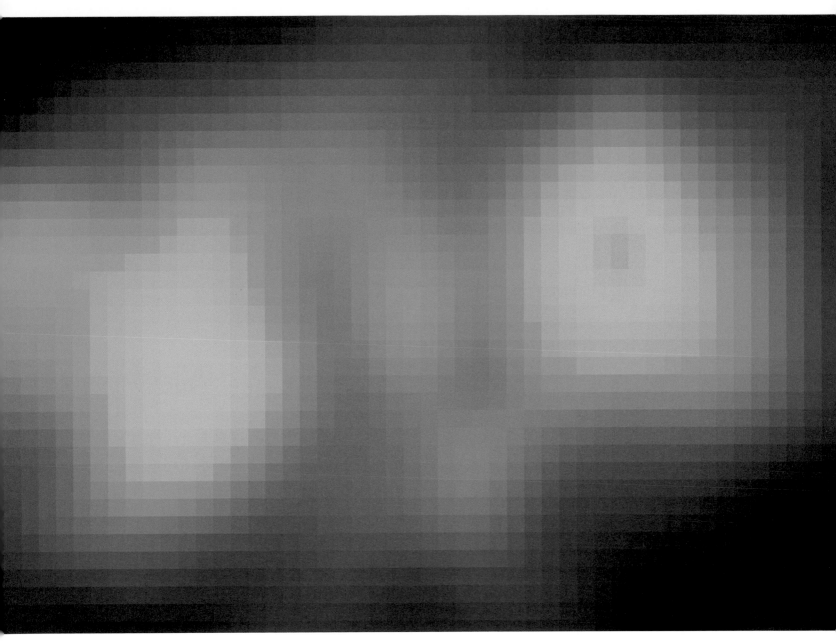

FIGURE 4.11. An x-ray binary system. This is a Chandra image of an x-ray binary system called SS 433 that contains a low-mass black hole and a star with a mass about 20 times that of our Sun. The yellow clumps are a gigantic traffic pileup of 50-million-degree gas blobs 5 trillion kilometers apart on opposite sides of the binary system. This gas is located at the ends of two narrow jets and is heated over and over again from collisions within the jets. The black hole in this system shoots blobs of material outward every few minutes like bullets. The blobs speed off at a quarter the speed of light and then a few months later they begin to pile up and smash into each other as they start to slow down.

FIGURE 4.12. Moving jets. This series of Chandra images of XTE J1550-564 from June 2000 to June 2002 shows the central black hole and the eastern and western jets emerging from it. Here we can watch the growth of the jets in this source. An x-ray outburst was detected in 1998 before these pictures were taken. Chandra saw the leftmost jet of high-energy particles appearing first, and then the rightmost jet. These jets moved away from the black hole at about half the speed of light. Four years after the original outburst, the jets had moved three light years apart and the left jet finally disappeared.

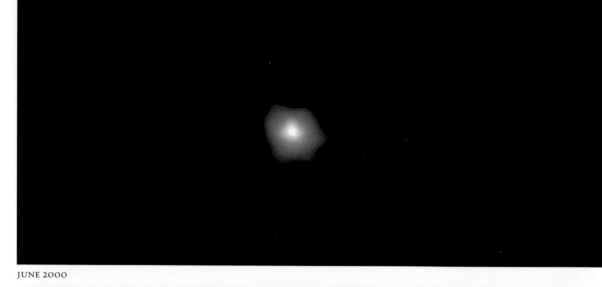

JUNE 2000

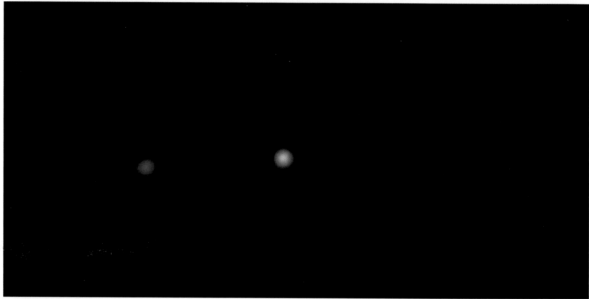

AUGUST 2000

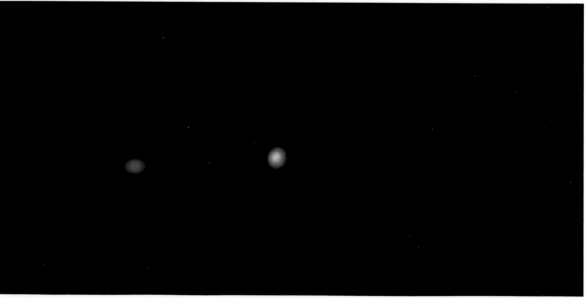

SEPTEMBER 2000

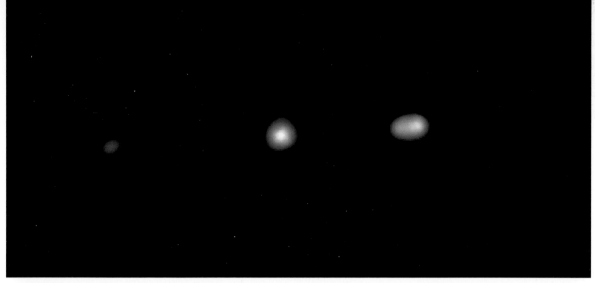

MARCH 2002

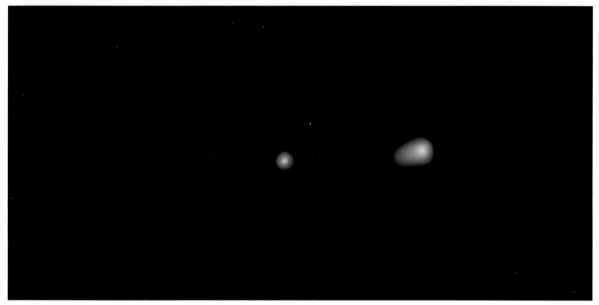

JUNE 2002

5

Fires in the Sky

E'VE SEEN that the Milky Way galaxy hosts a remarkable menagerie of powerful x-ray sources. But these fireworks pale in comparison to the amazing pyrotechnic displays in other galaxies, some of which contain dense regions of violent and rapid star birth and death, with stellar birthrates that are tens to hundreds of times higher than the birthrates in the Milky Way. Other galaxies have spawned giant, galaxy-sized "explosions" of enormous power that are sustained for many millions of years by the combination of intense radiation from massive young stars and the tremendous energy of multiple supernova blasts. Still other galaxies spend much of their lives within a cosmic bath of multimillion-degree gas. In all cases, the bizarre behavior of these galaxies is much easier to see in x-ray light than in visible light.

One beautiful example of a galaxy that is forming stars at an extremely rapid pace is named Messier 83 (shown in fig. 5.1). In typical fashion for a spiral-shaped galaxy, the visible-light image displays an exquisite pinwheel of stars, gas, and dust. The x-ray image, on the other hand, allows the regions of violent activity to shine through the gas and dust to reveal an entirely different picture. Here, the spiral arms contain many distinct and bright x-ray sources. These are stellar binary systems, just like those in the Milky Way, that contain neutron stars or black holes sucking material from their companion stars. The yellow, fuzzy patch of x-rays at the center of the galaxy is gas that has been heated to a temperature of 7,000,000°C. This central region is actually a giant stellar nursery, filled with massive stars that are being born rapidly and living shorter lives than the stars in our own galaxy. The tremendous burst of star formation that causes the center of M83 to glow so brightly began about 30 million years ago. Because the stars here are so massive and are evolving so quickly, there are about six supernova explosions per century, roughly three times the rate in our Milky Way. The fainter, fuzzy patches of x-ray light spread throughout the spiral arms in M83 let us know that star formation is happening there too, but at a much slower rate than toward the center.

Galaxy fireworks can also be caused by collisions or near collisions between galaxies. When two galaxies pass close to each other, their mutual gravitational attraction can wreak havoc on the stars, gas, and dust in both galaxies, with the smaller one usually suffering the most damage. Sometimes a near collision forces the gas in the spiral arms to flow inward and collect toward the center of one or both galaxies. Once enough gas collects near the center, it will start to collapse under its own weight and begin to form stars. Figure 5.2a shows a pair of galaxies of very different size that are currently interacting with each other. Their close proximity has likely caused the stellar activity represented by the bright x-ray patches in the galaxy cores (fig. 5.2b), especially in the case of the smaller galaxy, which is suffering an extreme case of gravitational stress from the larger galaxy. The small visible and x-ray pictures (figs. 5.2c and 2d) zoom in on the center of the larger galaxy, where there is something even more intense going on. Here, x-rays are made by material that is rushing outward from a massive black hole at a speed of approximately 1.5 million miles per hour (680 kilometers per second). This black hole is similar to the black hole at the center of the Milky Way, but it's not as silent. It's having a much greater impact on its neighborhood.

When two galaxies finally collide and begin to merge with each other, their fireworks become even more dramatic. Because of the severe gravitational disturbance to both galaxies, such collisions may cause recurring cycles of forming stars. As the galaxies orbit and are pulled toward each other, the stars and gas are repeatedly disrupted and torn away from their normal paths of motion. This affects more than just the center of each galaxy—clouds of dust and gas in the disks also start to collide and collapse to form new stars. Sometimes galaxies go through phases when millions of stars are being born, living their lives and exploding within a very short timespan. Astronomers call this phase a "starburst" phase. Because things are happening so quickly, a starburst doesn't last very long—only a few million years, compared to the entire lifetime of the galaxy, which might be tens of billions of years or more.

The Antennae pair of galaxies (fig. 5.3) is a striking example of a classic starburst, in this case produced by two colliding spiral galaxies. The Antennae received its name from the wide wispy streams of gas, dust, and stars resembling an insect's antennae (fig. 5.3a). These gigantic wisps, which are hundreds of thousands of light years across and have been ripped from the two galaxies by their mutual gravitational pull, were produced by the collision between the galaxies that began approximately 100 million years ago. The most striking features in the visible image (fig. 5.3b) are the extensive dust lanes that crisscross the galaxies and absorb much of the starlight from our view. In the x-ray image (fig. 5.3c), the dust lanes no longer block our view, and a galactic merger is revealed that is aglow with the fireworks of star formation. Most of the individual bright x-ray sources are normal x-ray binary systems, while the red fuzzy patches are x-rays erupting from bubbles of hot gas that have grown from the accumulated power of hundreds to thousands of exploding supernovae. Besides the normal x-ray binary systems, several of the pointlike x-ray sources are extremely bright, ranging from tens to several hundreds of times brighter than similar sources in our Milky Way. These objects are a puzzle and are perhaps an entirely new type of x-ray source, which is putting out much more power than is possible for normal x-ray binaries.

It is not entirely clear how galaxies evolve, but astronomers believe that colliding spiral galaxies like the Antennae will eventually merge to form a single elliptical galaxy. As gravitational forces tear two spiral galaxies apart, the organized structure of the stars and gas is eventually destroyed and the galaxies combine and transform themselves into what looks like a giant ball of stars. The galaxy NGC 4261 is an example of an elliptical galaxy that may have resulted from the merger of two galaxies, at least one of which was a spiral galaxy. In visible light (fig. 5.4a), the record of past spiral structure is all but lost, but in x-ray light (fig. 5.4b), we can see the fossil prints of a collision that some astronomers believe happened billions of years ago. The ribbons of x-ray sources that are strung across the top and bottom of the image suggest the pattern of x-ray binaries that once were stretched out in spiral arms. One idea is that a smaller galaxy strayed too close to and was captured and ripped apart by a much larger galaxy. As the gas was flung into tails tens of thousands of light years across (like the Antennae), the gas collapsed and formed massive stars, many of which were born as binary systems. This x-ray evidence will be seen for hundreds of millions of years—long after the large galaxy swallows up the small galaxy. X-rays may in fact be one of the best ways to detect past galactic mergers.

Starbursts can put on quite a show for our x-ray telescopes. After a starburst has been going strong for a few million years, the energy produced by all of the stellar explosions and stellar winds combines to create an enormous flow of energy called a superwind. Superwinds are so powerful that they can push themselves out of a galaxy at enormous speeds. The hot gas in a superwind glows spectacularly in x-ray light. Figure 5.5 shows a galaxy called NGC 3079, which has delicate filaments of gas that form the edges of a cone-shaped bubble just above the center of the galaxy. The bubble consists of streams of particles and hot gas that have burst through the disk of the galaxy at almost 2.2 million miles per hour (1,000 kilometers per second).

As this bubble of energy rips through the disk, it tears away cooler gas in the disk and carries it along for the ride. Because NGC 3079 is a fairly massive galaxy with a large gravitational pull, the gas will eventually slow down and fall back onto the disk. If we could speed up our view of NGC 3079 so that one hundred thousand years passed each second, we would see many bubbles popping up from the galaxy and the gas raining back down on the disk, much like a seething cauldron.

Sometimes an entire galaxy is involved in the starburst. Figure 5.6 shows the edge-on starburst galaxy NGC 4631. In visible light, the galaxy looks warped but otherwise fairly normal. However, we see a much clearer picture of the star formation in ultraviolet light, where the disk of the galaxy is enveloped in giant bursting bubbles created by clusters of massive hot stars. The x-ray image completes this picture by showing a giant radiance, or halo of hot gas, that has escaped the galaxy and now surrounds it. Figure 5.7 shows another galaxy that is surrounded by a halo of hot gas. This galaxy, named NGC 1569, has been forming stars for the last 10 to 20 million years. Its million-degree halo is due to the energy of hundreds of supernova explosions that have combined to push gas out of the galaxy. Because this galaxy is much smaller than NGC 3079 and NGC 4631, its gravity is not strong enough to hold on to all of this gas. So the gas will eventually escape the galaxy completely and carry elements formed inside stars, such as oxygen and iron, into intergalactic space. Small galaxies like NGC 1569 are the most common in the universe. They facilitate the recycling of elements created in stars, which would otherwise not be available to form new stars and planets.

Another gorgeous example of an outflowing superwind from a starburst is apparent in the nearby galaxy Messier 82 (fig. 5.8). Being only 11 million light years from Earth, M82 is one of the nearest starburst galaxies. Massive stars are forming and dying in M82 at a rate ten times higher than in our own galaxy. Astronomers believe that sometime in the last 100 million years, M82 had a close encounter with a neighboring galaxy called M81, which is ten times more massive and only one hundred thousand light years away. The gravitational pull of M81 probably stressed M82, causing its gas clouds to collapse to form stars and strongly altering its appearance. The bright points of x-ray light within M82 (fig. 5.8b) are supernova remnants and x-ray binaries, while the diffuse x-ray light in red extends across several thousand light years and represents multimillion-degree gas being pushed out of M82 at speeds of near 1 million miles per hour (500 kilometers per second). Violent episodes of star formation began this process, which will continue for millions of years.

Some of the very bright point sources in the Chandra image of M82 give off from ten to a hundred times more x-ray light than similar sources in our galaxy. In other words, they are more powerful than normal stellar binary systems that contain a neutron star or stellar-mass black hole pulling material from its companion star. Two reasons for this brightness seem most likely. They could be normal binary systems oriented in such a way that their light is concentrated in our direction, making them appear brighter than they really are. Another way to account for their brightness is that they may contain much more mass than normal x-ray binaries. The only way this could happen is if they contain mid-mass black holes, with masses that fall somewhere between low-mass black holes (two to ten times the mass of the Sun) and high-mass or supermassive black holes (at least a million times the mass of the Sun). Mid-mass black holes are an entirely new type of black hole that was only recently discovered by astronomers.

The most interesting x-ray source in M82 is a bright source near its center, but offset from the center of the galaxy by about 600 light years (fig. 5.9). This x-ray source brightens and dims over a period of several months—a pattern that suggests that it is indeed a mid-mass black hole. Such a black hole could form either from the collapse of an extremely large star that grew from the coalescence of many stars or from many smaller black holes coming together and merging into one.

Once mid-mass black holes have formed, it may be possible for them to migrate toward the center of a galaxy and merge further to form a supermassive black hole like that in our galaxy's center. In the nearby starburst galaxy NGC 253, the stage may be set for this process to occur. NGC 253 is located about 8 million light years from Earth and is the nearest starburst galaxy. The visible image (fig. 5.10a) shows a normal-looking galaxy, but the rate at which stars are forming in NGC 253 is fifty to one hundred times that for a normal galaxy its size. The starburst behavior has been going on for tens of millions of years. Since it only takes a few million years for a massive star to use up its fuel and explode, stars are now exploding at a rate of about one every five to ten years. These violent explosions heat the surrounding gas to produce expanding hot bubbles that blast out of the galaxy disk along the path of least resistance. The enlarged picture of the center of the galaxy (fig. 5.10b) is an intensity-colored x-ray image that shows a concentrated region of stars (the white area) and gas (the red area) that is left over from an episode of star formation that began almost 90 million years ago.

Of more than 50 bright x-ray point sources in NGC 253, there are a handful of powerful ones that may be mid-mass black holes. One of these is lurking within 45 light years of the galaxy's core and is buried behind a giant disk of dust and gas that is about 900 light years across. The shadow of this dusty disk shows up in the energy-colored x-ray image as a green strip across the center of the galaxy (fig. 5.10c). Since the mid-mass black hole is located so close to the center of the galaxy, it may be gravitating inward and, if so, could eventually grow into a supermassive black hole.

Supermassive black holes may actually occupy the centers of many starburst galaxies. Some are already known to possess them, like the largest galaxy of the M51 pair (fig. 5.2). But these black holes tend to be fairly modest. The majority of the energy in starbursts comes from the x-ray binaries and supernova remnants in the disks of the galaxies and the powerful, galaxy-sized superwinds. In terms of the total power that is produced in their centers, starburst galaxies are the poorer cousins of active galactic nuclei, which are examined in the next chapter. ✳

FIGURE 5.1. The Southern Pinwheel galaxy. Located 12 million light years away in the direction of the southern constellation Hydra, is the elegant galaxy Messier 83 (M83). M83 is a spiral galaxy, similar to our Milky Way, and is also called the Southern Pinwheel galaxy. It is named after the French astronomer Charles Messier, who first published its location in 1781. The visible image (a) shows details of the stars—a spherical central bulge teeming with old, yellow stars, and grand spiral arms containing bright clusters of young, blue stars, crisscrossed by patchy dust lanes. The energy-colored x-ray image (b) presents an entirely different picture. Neutron star and black hole binaries are revealed here as bright points of light, while the bulge of the galaxy glows with 7-million-degree gas from a concentrated burst of star formation that began about 30 million years ago. Slightly cooler gas is spread along the spiral arms in fuzzy patches—heated by widely separated supernova explosions and stellar winds. Bright, individual x-ray sources from the central star-burst are easiest to see in the enlarged image of the center of the galaxy (c), which is about 900 light years across.

◄ A

Pages 96–99

FIGURE 5.2. The Whirlpool galaxy. Messier 51, or the Whirlpool galaxy, is another exquisite and well-known spiral galaxy. Located about 30 million light years from Earth in the constellation Canes Venatici (hunting dogs), M51 was discovered in 1733. Its badly distorted and smaller companion galaxy wasn't recognized as a separate galaxy until 1781. Astronomers believe that the Whirlpool's distinct spiral structure is at least partly due to the gravitational effects of a close encounter with its smaller companion. As with M83, the visible image (a) shows the stars and gas, while the energy-colored x-ray image (b) shows x-ray binaries containing neutron stars and black holes surrounded by hot gas from star formation. But there is more going on here. The smaller galaxy is aglow in x-ray light because it is being torn and heated by the gravitational pull of the larger galaxy. Also, the center of M51 (c, d) is much brighter than the center of M83. The yellow, elongated cloud of x-ray gas (2,000 light years across) in the intensity-colored image has been heated to a temperature of 6,000,000°C by shocks caused by material that is flowing out of the center of the galaxy at a speed of 1.5 million miles per hour (680 kilometers per second). This powerful and concentrated outflow of energy suggests that a massive black hole is lurking there.

A

B ▶

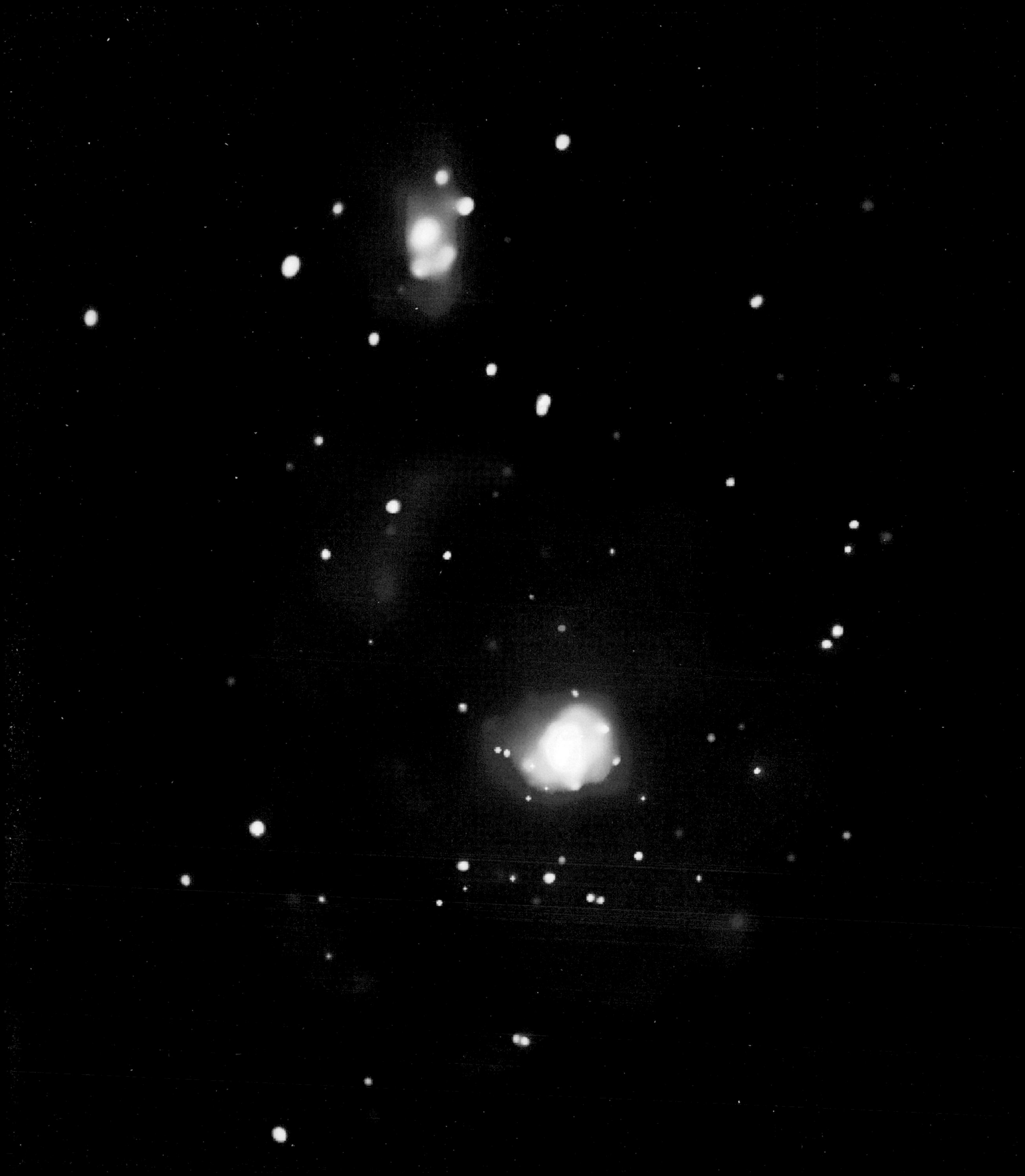

D ▶

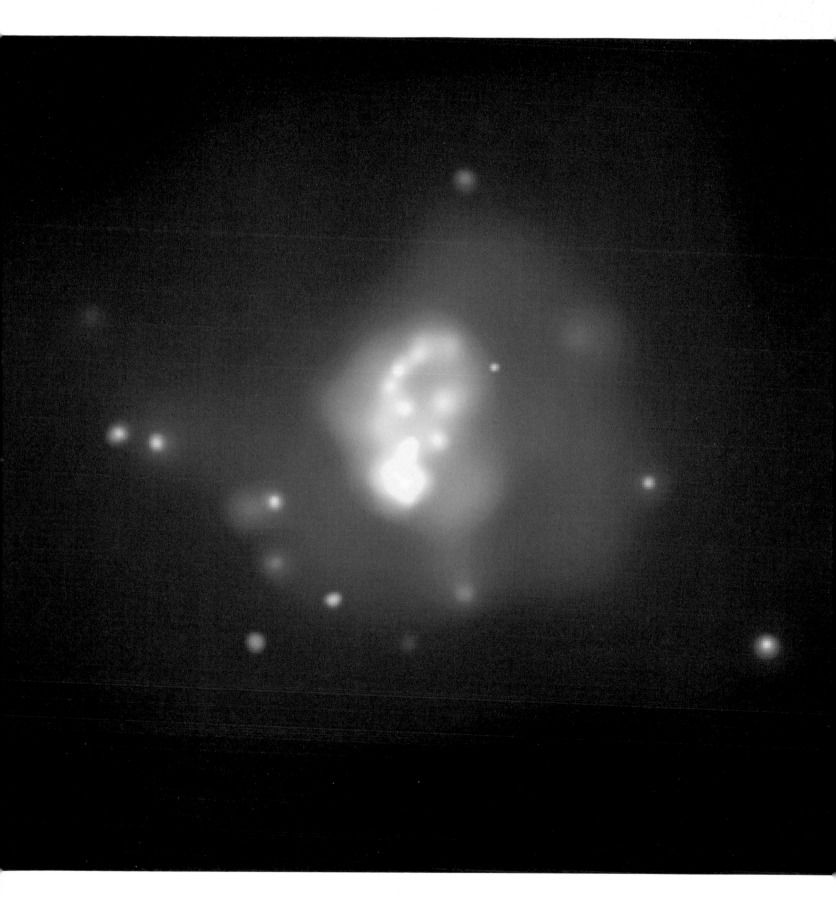

FIGURE 5.3. Cosmic collision. Sixty million light years distant in the constellation Corvus lies an impressive display of galactic fireworks called the Antennae. This pair of colliding galaxies (a) gets its name from the large wispy tails of stars, dust, and gas—resembling an insect's antennae—that have been torn away by the mutual gravitational pull of the galaxies. The bright orange patches of stars and dust seen close up in the HST image (b) are the individual nuclei of the two galaxies. There are over 1,000 bright young star clusters near the nuclei and in the spiral arms, which contain hot massive stars and active regions of star birth triggered by the cosmic collision. The reddened clouds of dust and gas are being compressed by the collision and will also eventually collapse and result in the birth of millions of new stars. In the energy-colored x-ray image (c), the dust lanes no longer block our view, and the bright points of x-ray light expose black holes and neutron stars. The red fuzzy patches are x-rays erupting from bubbles of hot gas that are thousands of light years across.

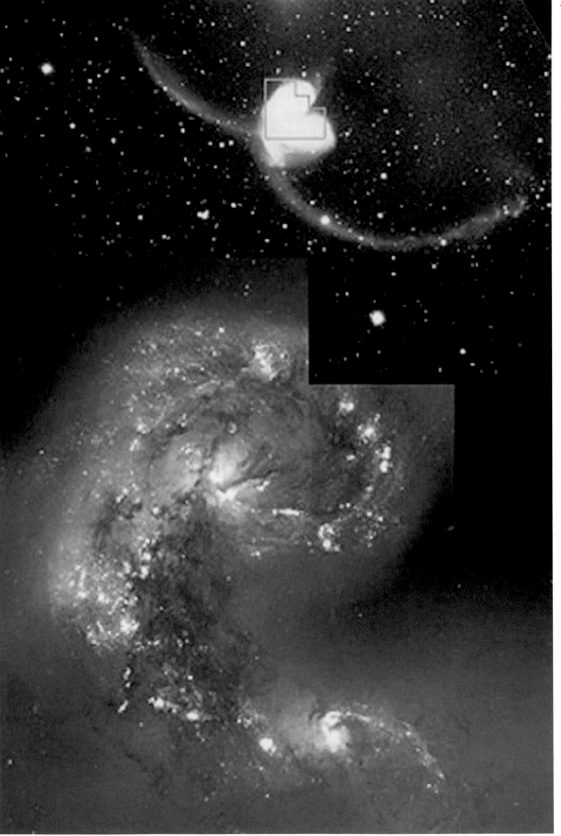

C ▶

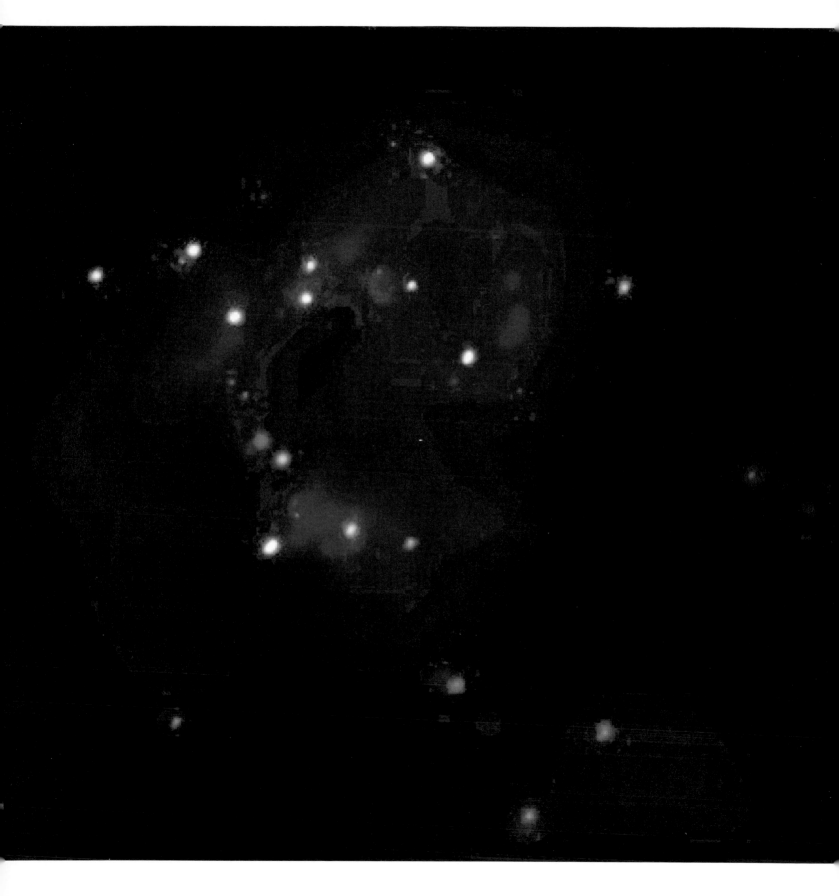

FIGURE 5.4. Fossil galactic remains. NGC 4261 is an elliptical galaxy 100 million light years away in the constellation Virgo. This giant ball of stars with very little visible gas and dust was probably once two galaxies (a). Astronomers believe that a large galaxy captured and swallowed a smaller spiral galaxy, which has left evidence of its past shape in the form of x-ray binaries that are strung out in ribbons. The x-ray image (b) shows the main galaxy—filled with hot gas at a temperature of 7,000,000°C—surrounded by bright x-ray points of light that appear to be arranged in the shape of elongated spiral arms that have been torn apart. This suggests that NGC 4261 once had extended wispy structures like those in the Antennae (fig. 5.3) and that, over time, the gas in those structures was used up to form stars. All that's left now, after hundreds of millions of years, are the long-lived x-ray binaries. In a sense, we can use x-rays to examine the scene of the crime and find traces of past activity, like detectives searching for evidence to convict a suspect.

A

B ➤

A

FIGURE 5.5. A seething cauldron. NGC 3079 in the constellation Ursa Major is a lovely edge-on spiral galaxy 55 million light years away. Gasses are escaping this galaxy much the way bubbles rise from a boiling cauldron. The bubble of gas in the central region is 3,000 light years across and signifies a strong starburst. The full image of the galaxy (a) shows heated x-ray gas (blue) together with cooler visible gas (red and green). The horseshoe-shaped feature in the enlarged image (b) is the brightened edge of a superwind that is blowing out of the galaxy at a speed of 2.2 million miles per hour (1,000 kilometers per second). The gas within this cone has a temperature that ranges from 10,000°C to about 10,000,000°C. As hot gas and energetic particles are rushing out of the galaxy, they drill a large hole and strip away streams of cooler gas.

B ➤

FIGURE 5.6. Galaxy halo. Located about 25 million light years from Earth in the constellation Canes Venatici is the spiral galaxy NGC 4631. This edge-on spiral is similar in size to the Milky Way and has a significant warp in its disk, making it look something like a swimming whale and earning it the nickname the Whale galaxy. The warping appears to be from the gravitational interaction with the small fuzzy companion galaxy seen in the image and another companion galaxy that is somewhat farther away. With the exception of its unusual warp, the galaxy looks normal in the visible image (a) with dust lanes and young, blue stars and old, yellow stars near the center. However, the x-ray image (b) shows a soft blue glow of x-ray light surrounding the disk from hot gas that is flowing out of the galaxy due to the combined effects of stellar winds and supernovae. The composite image (c) paints a complete picture in different wavelengths of light. Here the x-ray light (blue/purple) surrounds the swirling cauldron of star formation seen in ultraviolet light (faint orange).

A

B

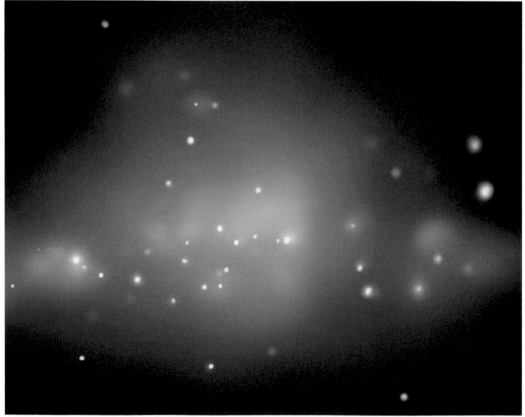

A

FIGURE 5.7. Escaping gravity. The galaxy NGC 1569 is much smaller than the Milky Way and is 7 million light years away in the constellation Camelopardus. The visible image (a) shows an irregularly shaped and uneventful-looking galaxy. But the x-ray image (b) shows that a huge amount of hot gas is escaping the gravity of the galaxy. This galaxy underwent a strong starburst about 10 to 20 million years ago. The supernova explosions from this starburst eject material into the gas and heat it to millions of degrees. The hot bubbles of gas then flow out of the galaxy at high speeds. Because this galaxy is small, its gravity is low, so its gas will eventually escape entirely and be lost in intergalactic space.

B ➤

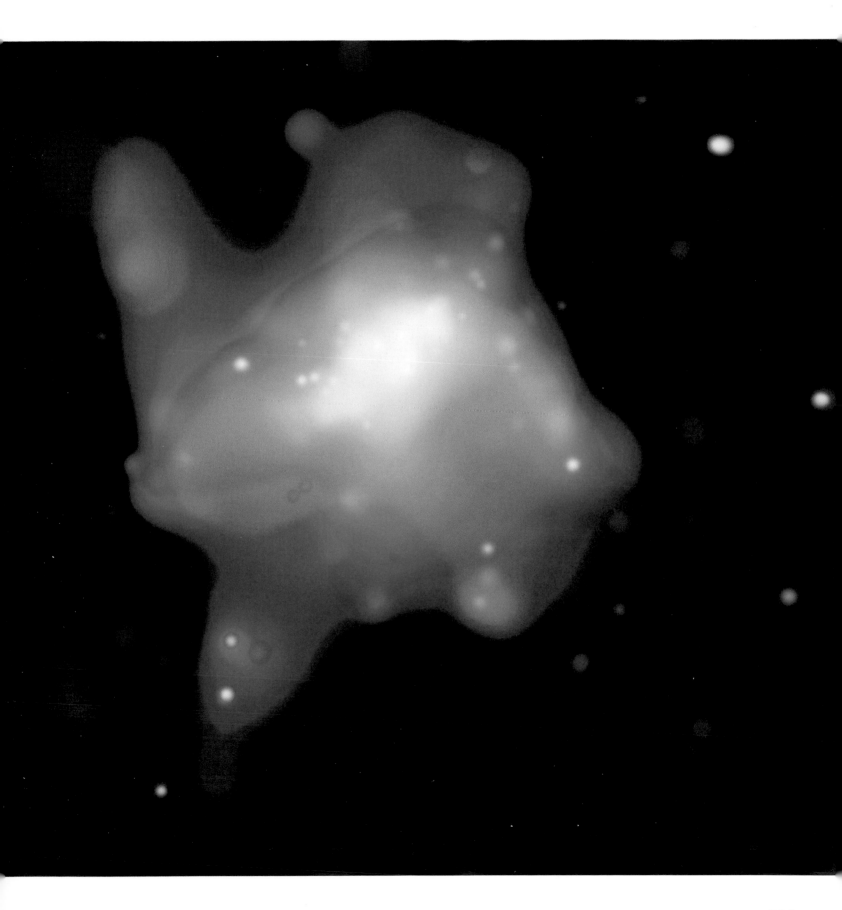

FIGURE 5.8. An exploding star-burst. The starburst galaxy Messier 82 is located 11 million light years from Earth in the constellation Ursa Major (Big Bear). M82 is most properly called an irregular galaxy and was first discovered in 1774. Nicknamed the Cigar galaxy, M82 was once thought to be an exploding galaxy, but we now know that it is merely a stunning example of a starburst galaxy, with its energy dominated by violent stellar behavior. The dramatic and delicate filaments in the visible light image (a) which make the galaxy appear to be exploding, are low-temperature gas escaping from the galaxy's core. Astronomers believe that a past encounter with a nearby galaxy is the cause of the fireworks in this galaxy. The x-ray image (b) shows hotter gas that surrounds the galaxy, as well as many x-ray binaries.

B

◄ A

Fires in the Sky

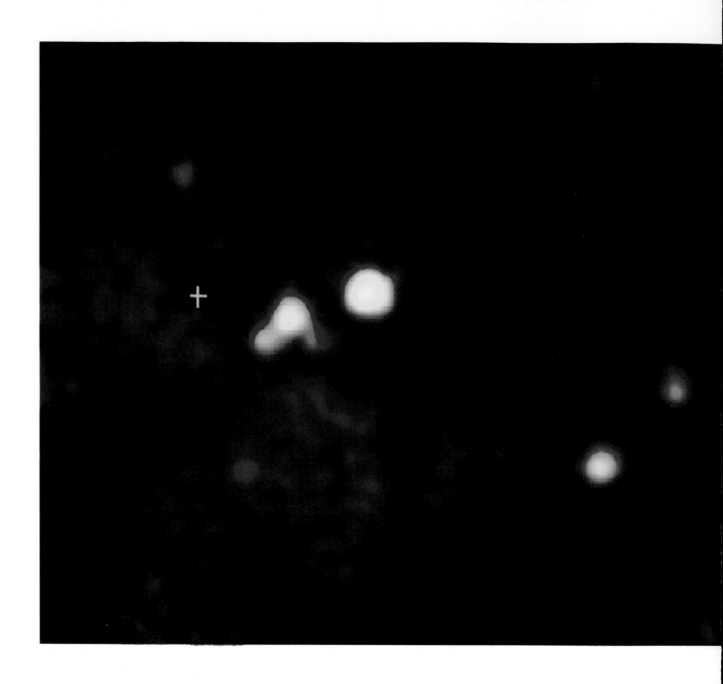

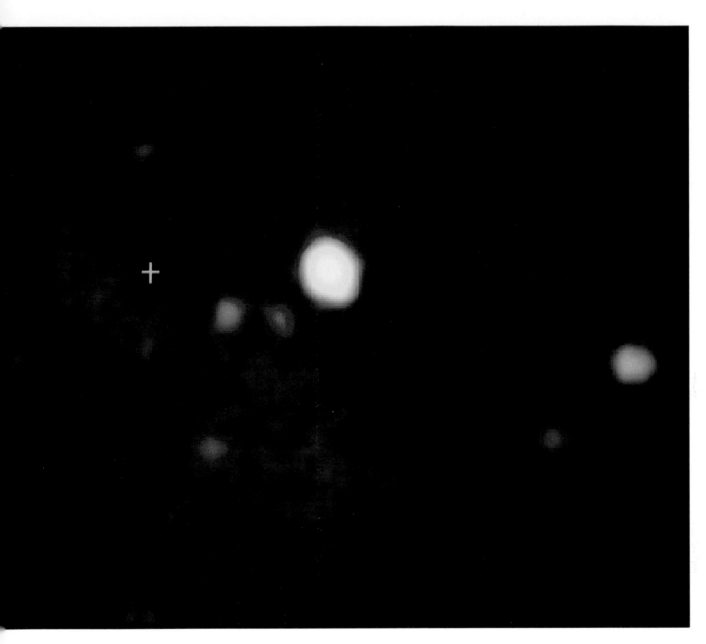

FIGURE 5.9. A mid-mass black hole? Chandra x-ray images of the central region of M82 taken six months apart. Here there are many normal x-ray binaries that contain neutron stars and stellar-sized black holes, as well as one unusual object that may be a mid-mass black hole. The x-ray brightness of the suspected mid-mass black hole (to the right of the green cross) varies and has faded in the image on the right. The large change in brightness in such a short time suggests that the x-rays are concentrated into a region that is so small that it can only be a black hole. The black hole is about 600 light years away from the center of its galaxy, which is marked by the green cross.

FIGURE 5.10. An evolving galaxy? The beautiful spiral galaxy NGC 253 is about the same size as our neighboring Andromeda galaxy (figs. 4.2 and 4.3) and is located 8 million light years away in the direction of the southern constellation Sculptor. Nicknamed the Silver Coin Galaxy due to its silvery-blue shimmer in visible images (a), NGC 253 was first discovered in 1783 by Caroline Herschel (sister of the famous astronomer William Herschel). NGC 253 is the nearest starburst galaxy to our Milky Way and contains violent star formation in its core. The intensity-colored x-ray image (b) shows the central region of NGC 253, several thousand light years across. The red area represents the outflowing hot gas from the central starburst. The energy-colored x-ray image (c) shows the shadow of a large dusty disk (in green) that surrounds the center of the galaxy. A mid-mass black hole appears to be lurking very near the center of the galaxy, buried behind this disk. This black hole may be gravitating toward the center, eventually growing to become a supermassive black hole.

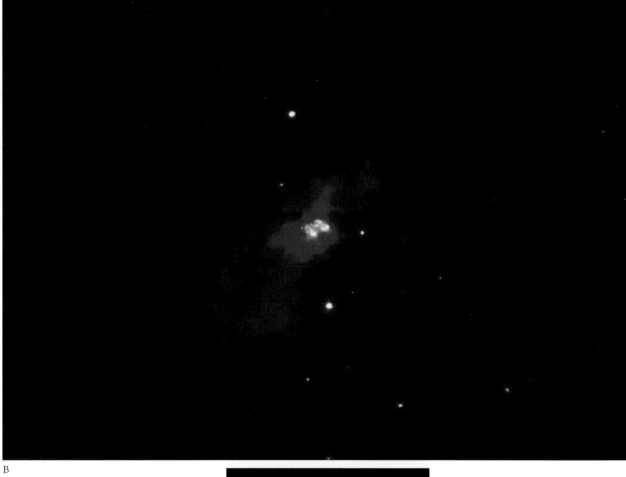

B

C

6

There Monsters Lie

BEHIND THEIR VAST and whirling veils of stars and gas, some galaxies contain incredible sources of energy buried deep within their cores. These mighty "central engines" are known to astronomers as active galactic nuclei, or AGNs, and the galaxies that harbor them are the kings of galactic nobility. AGNs can be as powerful as hundreds of billions of Suns or thousands of supernovae and they can effortlessly hurl streams of high-energy particles across hundreds of thousands of light years into intergalactic space. Such powerhouses are able to wreak havoc on a galaxy, and they often do.

The object at the heart of an AGN, the true monster, is thought to be a giant black hole that is several million to a billion times more massive than our Sun. There are several reasons behind this idea. First, it would be very difficult for anything other than a black hole to produce such large amounts of power. Normal events associated with stars, even supernovae, can't provide this much energy over a long period of time. Second, stars and gas at the centers of some galaxies such as the Milky Way orbit extremely rapidly, as noted in Chapter 4, which requires a highly concentrated gravity. Third, the light from AGNs tends to flicker on timescales of days to weeks, which means that the light must come from a compact area. Concentrating light in this way can happen when matter is packed as tightly as it is in a black hole—about the size of our solar system for a billion solar masses.

Astronomers believe that supermassive black holes probably exist in most, if not all, galaxies. But not all of these are AGNs, because like a spider that hides silently until an unlucky insect stumbles into its web, a black hole will only capture prey that wanders nearby. It doesn't go searching for a meal. Matter can either pass near a black hole and have its course altered, be trapped into orbit around the black hole, or plunge directly into the black hole. Since black holes gain mass and weight as material plunges beyond the event horizon, black holes will grow as they greedily "feed" from their surroundings.

Sometimes there is so much material that the black hole cannot devour it quickly enough. This material can build up into a disk, called an accretion disk, that orbits the black hole, providing a steady supply of fuel (fig. 6.1). The material in the accretion disk may be torn from neighboring stars or it may fall into the center of the galaxy from far away (fig. 4.9). As matter falls toward the black hole, it is heated to millions of degrees and glows in x-rays. Astronomers detect and study black holes by searching for and capturing this x-ray light. Galaxies like NGC 7052 (fig. 6.2) have gigantic rings of dust and gas that are thousands of light years across and contain up to a million solar masses of material. This material may eventually sink toward the center of the galaxy and become fuel for its black hole.

Once a supermassive black hole begins to feed from its accretion disk, it becomes a full-blown AGN. Figure 6.3 shows what such an AGN may look like. Besides fostering violent conditions of extreme gravity, temperature, pressure, and density, supermassive black holes are capable of launching powerful, opposing jets of high-energy particles that stream outward at speeds close to the speed of light. The jets are produced when the gas that is falling toward the black hole is redirected by the strong forces that exist near the black hole. Such forces are so mighty that they pull the gas and magnetic fields away from the black hole and point them in the opposite direction, kicking material out at high speeds in narrowly directed jets, which are also a strong source of x-ray radiation.

The galaxy known as Centaurus A (figs. 6.4 and 6.5) is a classic example of an AGN with a supermassive black hole at its center. Centaurus A is one of the nearest AGNs and is about 11 million light years from the Milky Way. Far from being a normal galaxy, Centaurus A is a beautiful example of galactic cannibalism in progress. About 200 million years ago, astronomers believe, the large elliptical galaxy collided with a spiral galaxy, which can now be seen as the dark band across the center of the elliptical galaxy in the visible image (fig. 6.4a). A closer look at this dark band (fig. 6.5) reveals intricate lanes of dust and gas, which are actually the remains of the former spiral galaxy that was ripped asunder in the massive collision between the two galaxies. The AGN lies well hidden behind this dense veil of gas and dust.

In the radio and x-ray images of Centaurus A (fig. 6.4b and c), the AGN is apparent as the bright pointlike source of light at the center, as well as the strong jet of high-energy particles that is emerging from the region near the black hole. The jet extends about 12,000 light years from the nucleus of the galaxy. Near its center, it is blasting away from the black hole at nearly one-half the speed of light. On the side of the galaxy that is closest to us, the narrow inner part of the jet is easiest to see in the x-ray image, but at the end of the jet, the radio emission is brighter. Here, the jet fans outward after slamming into the surrounding gas. Without both of these pictures, we would not know the true size or shape of the jet. Combining the visible, radio, and x-ray images (fig. 6.4d) provides a powerful and complete picture of Centaurus A.

This galaxy also appears to have undergone a series of explosions, possibly emanating from the black hole at its center. The large blue arcs in the x-ray image consist of gas at a temperature of millions of degrees, which forms a ring that is 25,000 light years across. The ring's size and location suggests that it was spawned during a gigantic explosion in the nucleus of the galaxy about 10 million years ago. Astronomers believe that the AGN may have been responsible for this explosion. When the two original galaxies began to collide, material plunging into the center of the larger galaxy may have caused the already present black hole to grow to its current supermassive size. Soon after the AGN was born, in other words, when the black hole "engine" finally turned on, the dramatic release of energy from the AGN may have launched the opposing jets and produced the explosion that gave birth to the large ring of hot gas.

Jets are important because they can be used to guess some things about black holes that can't be seen directly. How jets are formed is still not well understood, but not all jets are alike, and this in itself tells astronomers something about the different ways they form. The next four sets of pictures show different types of jets. The first pair of images (fig. 6.6) depicts a quasar called 3C 273, which has a high-powered jet that is about 100,000 light years across. The jet is much larger than the one in Centaurus A, but in this case, the jet appears to only be tenuously connected to the AGN, which is the bright source of light to the upper left of the x-ray image. The lack of a connection between the jet and the AGN is only an "optical illusion," because there is material located between the x-ray jet and the AGN that is not giving off x-rays. After apparently being violently ejected from the core of the galaxy, 3C 273's jet eventually runs into dense material, causing it to suddenly slow down and pile up. As the jet slams into this dense gas, it heats the gas and creates x-rays.

The second set of images (fig. 6.7) shows a giant elliptical galaxy called M87, similar to Centaurus A but 50 million light years away. In this case, the jet has a beaded structure and is about the same size and shape in all of the images shown here. The bright galactic nucleus that harbors the supermassive black hole shines at the extreme left of the x-ray, visible, and radio images. The interesting thing about this jet is that the bright spots in the x-ray, visible, and radio images are at slightly different locations. Inside the jet, there are high-energy electrons that are accelerated by shock waves. These electrons spiral around magnetic fields and radiate photons of different types of light, creating different patterns in the radio, visible, and x-ray light. The difference in the location of the knots may be because the particles producing the radio light, for example, may live longer than the particles producing the x-ray light.

The next AGN, called Cygnus A (fig. 6.8), is one of the brightest radio sources in the sky, with beautiful radio jets that are as much as 600,000 light years across. But this radio galaxy is also bathed in a pool of hot x-ray gas. On their way out, the powerful radio jets produced by the central black hole have carved out an immense cavity in the gas, which is shaped like a giant football. The hot gas piles up into swirling ridges of x-ray light. Like the dense material that stops the 3C 273 jet, the hot gas provides enough resistance to slow down the particle jets, which end in bright spots 300,000 light years from the center of the galaxy.

Another very different jet is produced in the galaxy Pictor A (fig. 6.9). In the x-ray image, this jet is about 360,000 light years long and points toward a bright x-ray blob at the right of the image. This blob, or "hot spot" is 800,000 light years away from the origin of the jet, which is a total distance that is eight times longer than the Milky Way galaxy. The hot spot is where the jet is plowing into invisible gas in intergalactic space. One reason that the inner part of the jet is smoother than the knotty jet in M87 is because there is nothing for the jet to run into along its entire length. The hot spot is different. Where the jet ends, it runs into dense gas and creates shock waves, which are kicking electrons and protons up to speeds approaching the speed of light.

Not all AGNs produce jets, but they can be interesting in other ways. The beautiful, butterfly shape of the galaxy NGC 6240 (fig. 6.10) is the product of a collision between two smaller galaxies that happened about 30 million years ago. In addition to containing an AGN, NGC 6240 is also a starburst galaxy in which stars are forming, evolving, and exploding at a rapid rate. Heat generated by the starburst has created the large amount of hot gas in the x-ray image that gives the galaxy its trademark shape. But that's not all. Remarkably, NGC 6240 also has at its center not one but two giant black holes that are strong sources of x-rays. In the next few hundred million years, these giant black holes, which are about 3,000 light years apart right now, will drift toward each other and eventually merge to form a single, supermassive black hole. By merging in this way, black holes can combine forces to grow to enormous masses. This is one of the ways in which supermassive black holes may in fact be created.

So far we have illustrated two of the remarkable footprints of AGN activity: supermassive black holes and galaxy-sized jets. In the next chapter, we broaden our perspective from individual galaxies to galactic neighborhoods and the ways in which the galaxies interact with one another and influence their environment.

FIGURE 6.1. Chaotic motion. This drawing envisions a close-up view of an active galactic nucleus (AGN), with a supermassive black hole at its center. The disk of material rapidly orbiting the black hole contains dust and gas that has fallen inward from the surrounding galaxy and stars that have been torn out of their normal orbits. This disk is called an accretion disk and it provides all of the fuel the black hole needs to power jets of high-speed particles. As matter falls into the black hole, it uses its extreme gravity to convert some of this matter into enormous amounts of energy.

FIGURE 6.2. A giant ring of stellar gas. Telescopes cannot yet resolve accretion disks like the one imagined in figure 6.1 because they are very small—no more than a few hundred times the size of our solar system. But it is possible to see larger rings of gas such as the one shown here in the galaxy NGC 7052. This gigantic ring is about 4,000 light years across and contains more than a million Suns worth of material swirling around the center of the galaxy at speeds up to 340,000 miles per hour (150 kilometers per second). The black hole at the center of this galaxy is 300 million times more massive than our Sun. As the material from the ring gravitates slowly inward over a few million years, the black hole will have plenty of fuel.

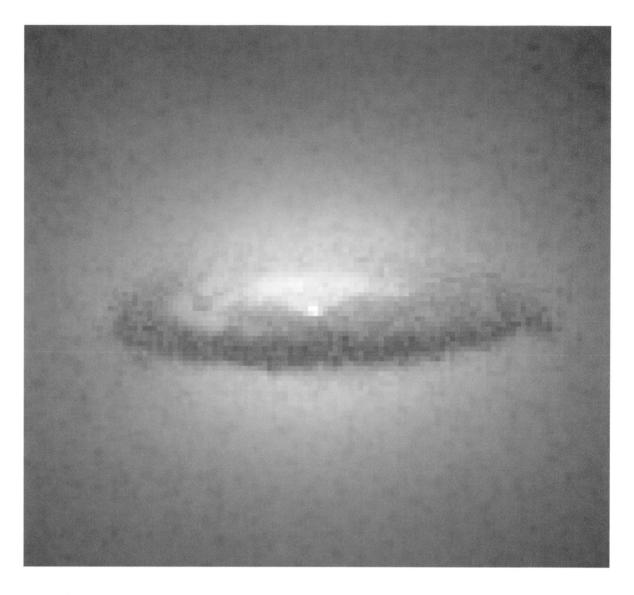

FIGURE 6.3 (opposite). An active galactic nucleus. Zooming far away from the black hole and spying on a galaxy from a distance, we can see exactly how a black hole is capable of dominating the scene. The tremendous jets, which have their source very near the black hole, may grow to be many times the size of the galaxy. This artist's concept shows a simplified version of high-speed jets powered by an AGN. As discussed throughout this chapter, however, real jets are rarely this simple.

A

Pages 125—128

FIGURE 6.4. Different views of an active galactic nucleus. Centaurus A is a galaxy 11 million light years away from Earth in the direction of the constellation Centaurus. It contains 300 billion stars and is classed as an elliptical galaxy, because it resembles an enormous ball of stars in visible light (a), but the fact that it is crossed by a galaxy-sized dust lane makes it very peculiar compared to most other elliptical galaxies. The AGN is not obvious in visible light but shows up clearly in the radio and x-ray images (b, c) as a bright point of light plus enormous jets that are speeding away from the black hole at half the speed of light. The x-ray picture also shows a vast ring that is about 25,000 light years across and angled 45 degrees away from us. The explosion that created this ring would have happened about 10 million years ago. When the visible, radio, and x-ray images are put together in figure 6.4d, the bizarre portrait of Centaurus A is complete.

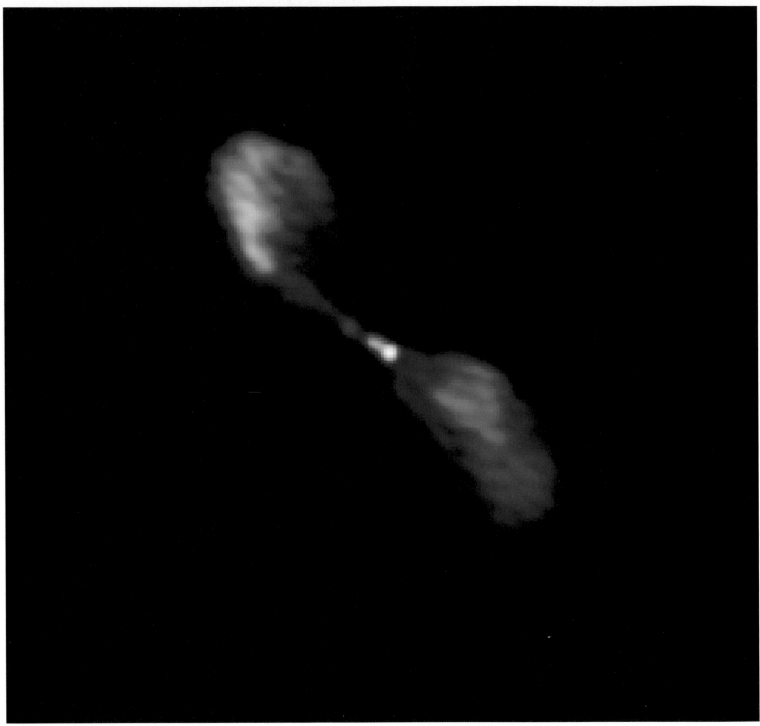

B

126

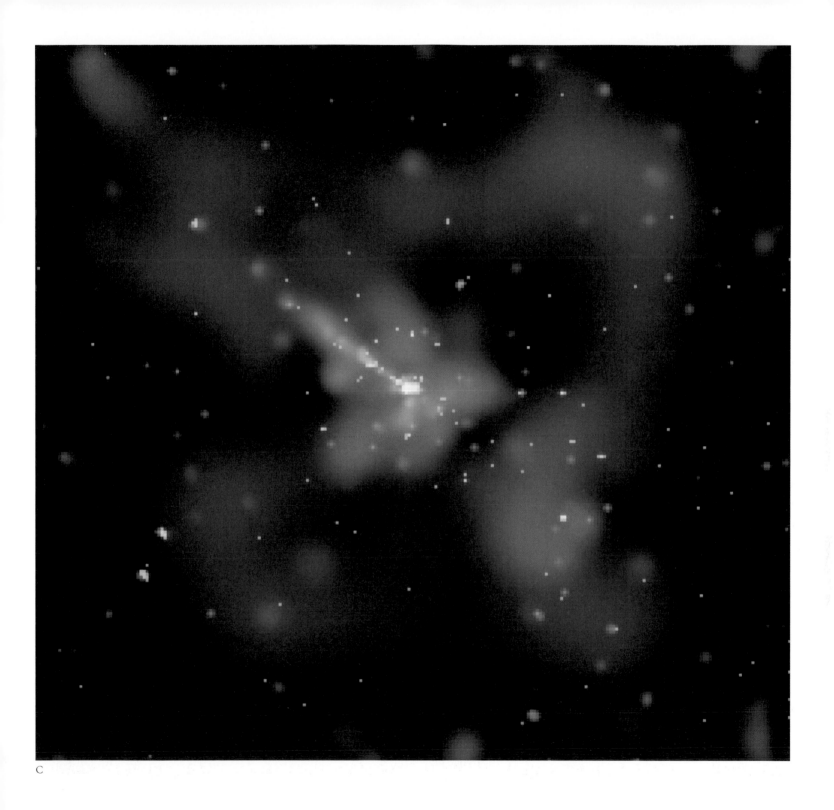

C

FIGURE 6.5. Galactic cannibalism. This visible image from the Hubble Space Telescope shows the complex central region of Centaurus A. The disturbed lane of dust crossing the galaxy is most likely all that is left of a spiral galaxy that was captured and cannibalized between 200 and 400 million years ago. The AGN is well buried behind this veil of dust, and so we don't see it in visible light. Although galactic cannibalism is only one theory of what happened to Centaurus A, this theory is well supported. Other recent images show the elliptical galaxy circled by a colossal wreath of stars that were torn out of the smaller spiral galaxy.

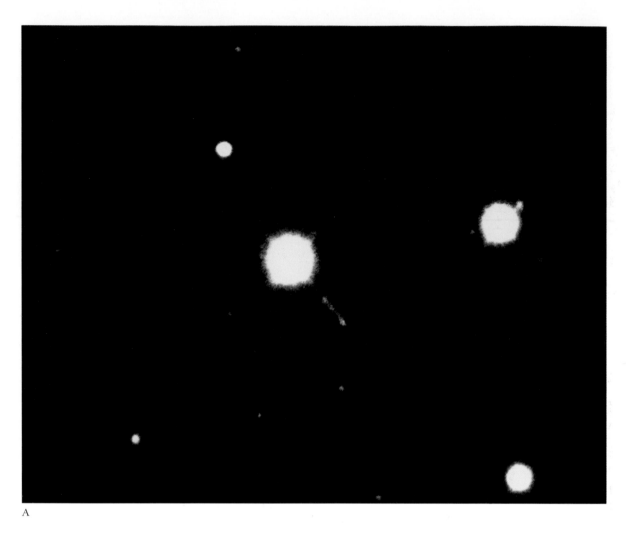

A

FIGURE 6.6. A distant quasar. The active galaxy 3C 273 is a quasar, or "quasi-stellar radio source." Quasars are bright radio sources associated with starlike objects and were first discovered in the 1960s. Astronomers initially believed that quasars were stars, but when the first quasar spectra were measured, it became clear that they were much too far away to be stars in our own galaxy and were most likely galaxies far away from us. The light in 3C 273 is redshifted by about 16 percent, which means that it appears to recede from us at a speed of 48,000 kilometers per second, placing it 3 billion light years away. The visible image shows a very bright galaxy with a jet 100,000 light years across (a). The jet is easier to see in the intensity-colored x-ray image (b).

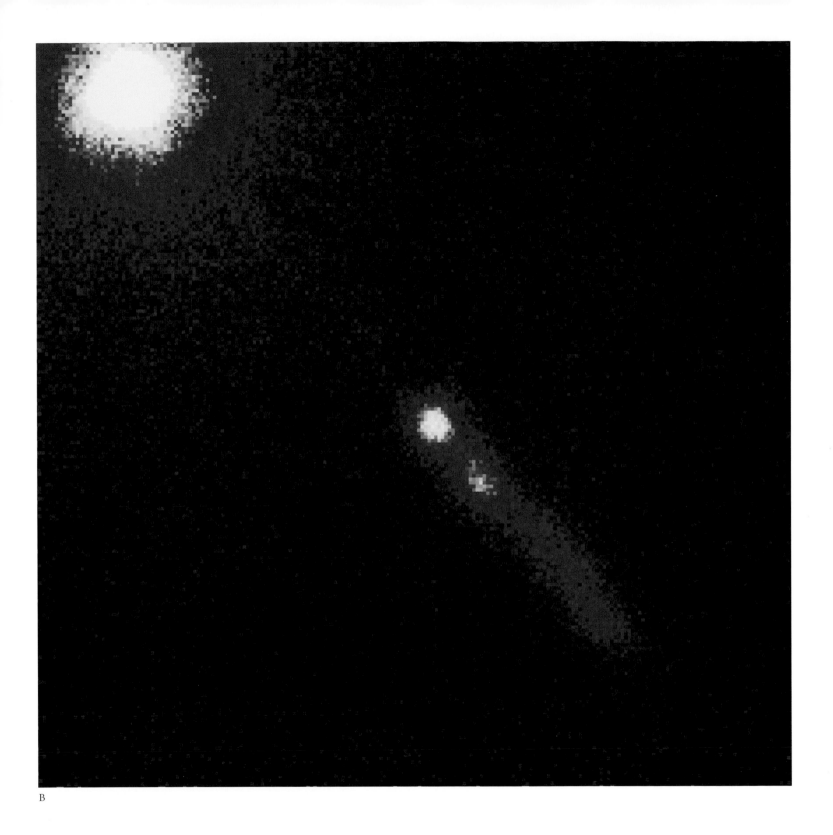

B

A

FIGURE 6.7. Radio galaxy. The AGN Messier 87 is a particularly bright radio source. The galaxy has an elliptical shape in visible light (a) and is located 50 million light years away in the direction of the constellation Virgo. The close-up pictures show the visible (b), radio (c), and intensity-colored x-ray (d) images of the jet, which is about 8,000 light years in length. There is very strong evidence in this galaxy for a supermassive black hole 3 billion times more massive than our Sun. Gas orbiting a mere 60 light years from the center is speeding around at 1.2 million miles per hour (550 kilometers per second).

B
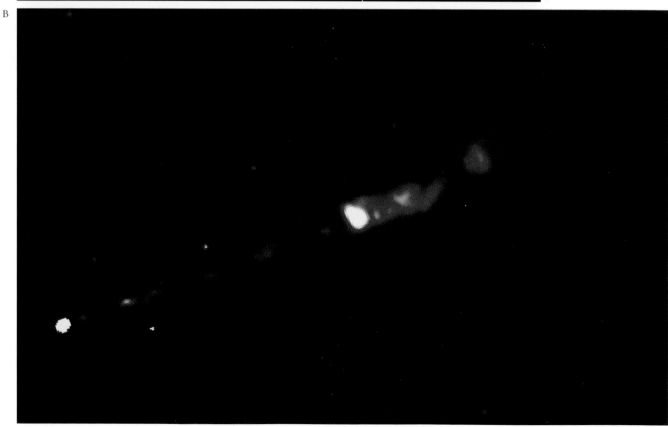

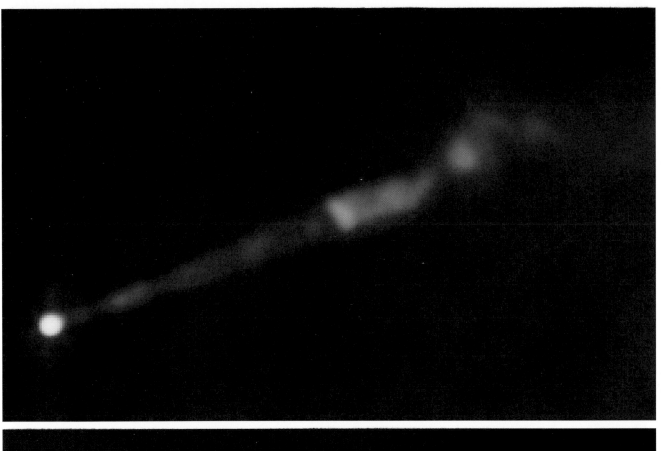

There Monsters Lie

A

FIGURE 6.8. Hot spots. The radio galaxy Cygnus A is located 700 million light years away in the constellation Cygnus (The Swan) and is one of the brightest radio sources in the sky. The visible image (a) shows a close-up view of the galaxy, where the supermassive black hole is buried behind dense gas and dust at the center. The radio image (b) shows the large and powerful jets from the black hole, which extend more than 600,000 light years across. The jets start out very narrow, and the supermassive black hole is believed to be the source of energetic particles rushing outward at near light speed. The giant lobes at the ends of the jets occur because the jets are slowed down as they plow into the gas that surrounds the galaxy. The intensity-colored x-ray image (c) shows an area of red-orange hot gas. The radio jets have carved out a giant football-shaped cavity. The x-ray image also shows bright hot spots where the jets abruptly plow into the cooler, dense gas that surrounds the galaxy.

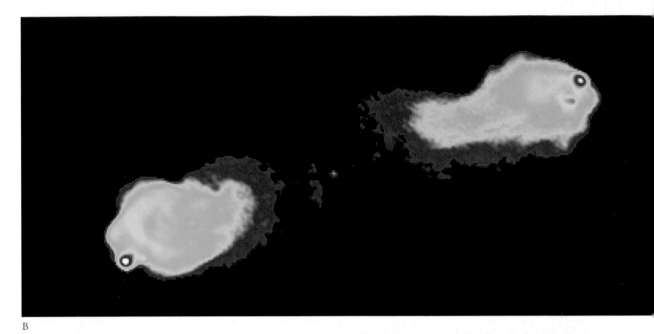

B

C

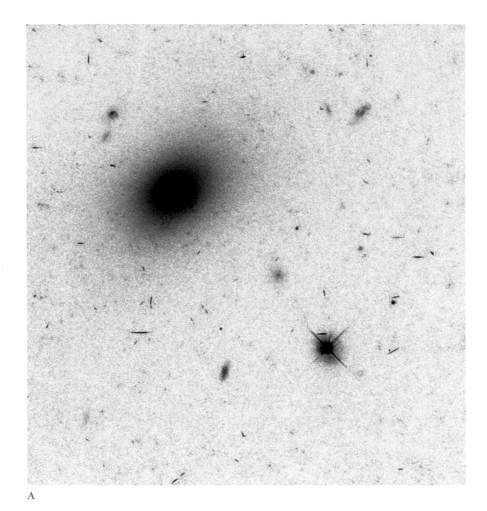

FIGURE 6.9. A smoking gun. The galaxy Pictor A, seen here in visible light (a), has a narrow jet that is nearly 800,000 light years across and made up of high-energy electrons and protons. The intensity-colored x-ray image (b) shows the glowing hot spot at the tip of the jet where it is piercing into a cloud of gas in intergalactic space. The distance between the origin of the jet and the hot spot is huge, more than eight times the diameter of the Milky Way.

A

B

FIGURE 6.10. Dueling black holes. NGC 6240, a butterfly-shaped galactic merger, is located 400 million light years away in the constellation Ophiuchus. The visible image taken with the Hubble Space Telescope (a) shows a very disturbed shape resulting from the violent collision of two galaxies. The red region in the energy-colored x-ray image (b) is multimillion-degree gas associated with the birth of stars resulting from the merger. Not one, but two giant black holes generate hard x-rays (the two blue points). When the hard x-ray image (blue) is combined with a blowup of the visible image (c), we see how the black holes relate to the merger. In a few hundred million years, these black holes may drift together under the influence of mutual gravity and combine to create a single supermassive black hole.

A

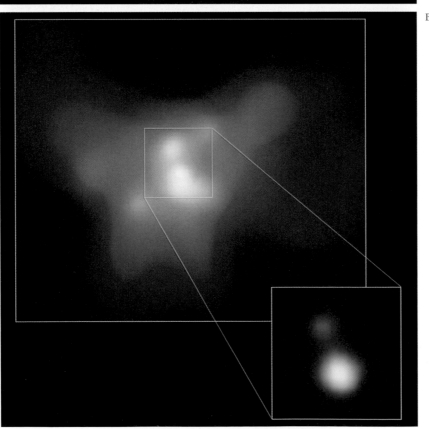

B

C

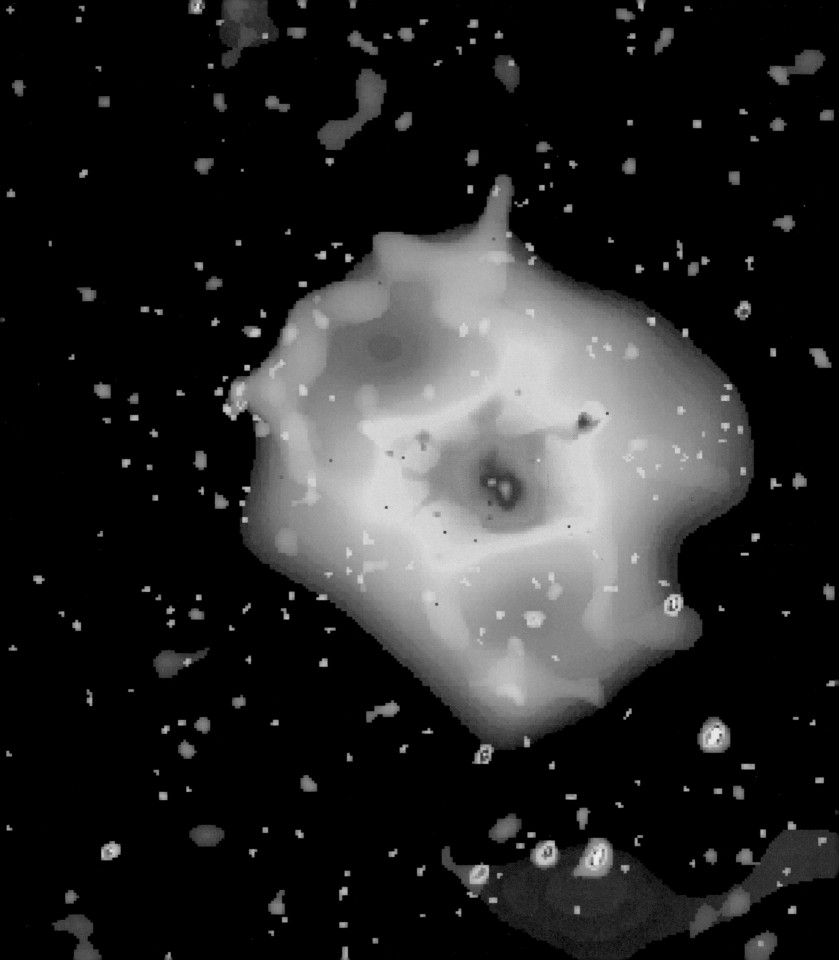

Cosmic Pools of Energy

7

WE LIKE TO THINK of galaxies as vast islands of stars floating majestically through space, but they rarely travel alone or even in pairs. Most galaxies live in bustling cosmic neighborhoods, bound to their lifelong companions by gravity. Small neighborhoods, or groups, as they are called, contain only a handful of galaxies. Groups can be as vast as several million light years across and may contain as many as 50 trillion stars. Large neighborhoods, or clusters, contain hundreds or even thousands of galaxies. Clusters can be over a billion light years across and may include as many as a thousand trillion stars.

Clusters of galaxies are by far the largest and most massive objects in the universe. They represent the effects of gravity on an immense scale. Since a cluster is held together by gravitational forces, the size and shape of a cluster depends on how much matter it contains and how this matter is spread out, in other words, on where most of the gravity is found. The appearance of a cluster also depends on its age. Astronomers believe that giant clusters have grown from smaller pieces, with galaxies coming together to make groups, and groups coming together to make clusters. The whole process began when the universe was young. Tiny concentrations grew in the smooth expanse of matter left over from the Big Bang. As these concentrations grew, their gravity attracted more matter, which made gravity stronger, which in turn attracted *more* matter in an unending cycle. The growth of clusters is therefore intimately related to the growth of the universe. The behavior of clusters as we see them today can help astronomers predict the ultimate fate of the universe.

A cluster of galaxies is an amazing sight. Figure 7.1a shows the Coma cluster of galaxies, which is located about 370 million light years from Earth. This dense cluster contains thousands of galaxies, and every galaxy harbors billions of stars. In visible light, Coma appears to be simply an immense collection of galaxies separated by vast distances. But far from being empty, the space between the galaxies contains an enormous pool of extremely hot gas (fig. 7.1b), with a temperature of about 100,000,000°C, in which the galaxies swim. The gas was heated long ago when it was captured and squeezed by gravity as the cluster first began to form. Gravity still plays a considerable role here. The small x-ray cloud to the lower right of the cluster is a group of galaxies that is being pulled into the cluster and will eventually become part of it. This is one way clusters continue to grow over time.

The x-ray gas in clusters reveals much more than how clusters are growing. We can also learn about the behavior of the galaxies themselves and how they affect the space around them. Figure 7.2 shows a close-up view of the center of the Coma cluster, where the x-ray light is brightest. The two white circles coincide with the two brightest galaxies in the visible image, indicating clumps of material with a much lower temperature than the surrounding gas. This material was probably ejected violently by stars and pushed out of the galaxies over the course of about a billion years. Normally, such clouds would have evaporated long ago, but the cool clumps appear to be in a delicate balance with the hot gas in the cluster.

Although they are less spectacular than clusters, galaxy groups are interesting in their own right, because they are the necessary building blocks of clusters. Figure 7.3 shows a famous grouping called Stephan's Quintet, named after the French astronomer E. M. Stephan who discovered it in 1877. If the galaxies look familiar to you, it's probably because they were made famous in the opening sequence of the 1946 Frank Capra film *It's A Wonderful Life*. The speeds at which these galaxies are traveling through space indicate that the group is located 280 million light years from Earth. It should be noted that not all of the galaxies Stephan discovered are true members of the group. His original quintet contained the five galaxies near the center of the image. But the large, bluish spiral galaxy is in fact much closer to us—only 40 million light years from Earth. This means that the original quintet is simply an accident of perspective. Luckily, the galaxy to the far left is also 280 million miles away from Earth, so the group has now become a quintet again.

The distorted shapes of the galaxies in Stephan's Quintet provide clear evidence that galaxies bound together by gravity can have an extreme effect on each other. The three galaxies in the center appear ready to smash together, like a giant cosmic car crash. The violent interactions are distorting the galaxies, tearing away trailing tails of stars and gas. More evidence for these violent forces can be seen in figure 7.4, which shows the Chandra x-ray image (blue) placed on top of the visible image. The space between the galaxies is partly filled with hot gas at a temperature of 6,000,000°C. This telltale gas has been stripped from the galaxies and then heated by the rapid movements of the galaxies, especially the spiral galaxy to the right of center, which is slamming into the gas at a speed of 2.2 million miles per hour (1,000 kilometers per second).

Because galaxy groups contain much less hot gas than their larger cluster cousins, the overall shape of the gas is very sensitive to the movements and behavior of individual galaxies. The gas in Stephan's Quintet clearly has a distorted shape. Another small group of galaxies called HCG 62 is shown in figure 7.5. Here the hot gas is much more clumpy and uneven than the gas in the Coma cluster. The two rounded features on opposite sides of the central galaxy (upper left and bottom right) are giant holes, or cavities, 30,000 light years across.

Astronomers believe these cavities were excavated by flows of high-energy particles produced by the central galaxy. If this were a large cluster with much more hot gas, it would probably be difficult to see such effects from a galaxy this size.

One important fact about clusters is that there is more mass contained in the hot gas than in all of the galaxies combined—about four to five times more. Furthermore, in order for a cluster to remain bound together so that it doesn't fly apart, it actually needs ten to fifty times more gravity than is supplied by the mass of the galaxies. Since the hot gas cannot provide this much gravity, some other form of matter must be providing the balance. This mysterious matter has been called dark matter because it cannot be seen at any known wavelength of light. It is the combined gravity of the galaxies plus the dark matter that holds the cluster together and keeps the hot gas from flying off into space. By measuring the amount of gas with our x-ray telescopes and measuring the amount of matter in the galaxies with our optical telescopes, we can determine how much dark matter is needed to keep the hot gas from flying away. The cluster Abell 2029, which is made up of thousands of galaxies (fig. 7.6) requires the equivalent mass of more than a hundred trillion Suns in mysterious dark matter to provide enough gravitational pull to keep its x-ray gas attached to the cluster.

Depending on their age, some clusters are better behaved than others. In young clusters, the galaxies are often disturbed and the gas can appear lumpy because it hasn't had time to mix thoroughly and settle down. In older clusters, the galaxies have stopped their chaotic motion and the hot gas fills the cluster in a regular pattern. The Coma cluster, for example, is a mature cluster. It has a round shape and the rings of color in the x-ray image (fig. 7.1b) show a consistent pattern toward the center of the image, which indicates a smooth shape throughout. In less mature clusters, the galaxies are still performing all sorts of acrobatics, from swooshing through the cluster at incredibly high speeds to smashing into each other. Measuring how fast the galaxies are moving and in what direction tells us something about how the cluster was formed. We can detect the movement of galaxies by looking for unusual features in the x-ray images or rapid temperature changes, which suggest shock fronts or collisions between galaxies.

Although you wouldn't know it from the visible image, the cluster Abell 2142 is in turmoil. In the visible image (fig. 7.7a), the galaxies look as though they are behaving normally. But this cluster actually contains two smaller clusters that are currently colliding—the main galaxies in these two clusters are indicated with arrows. Figure 7.7b shows the x-ray image of the cluster. The bright elliptical area lies exactly where the two main galaxies are located, marking the spot where the two small clusters are smashing together. The temperature of the gas in this spot is about 50,000,000°C and is cooler than the gas in the larger cluster. This region behaves like a giant atmospheric pressure front, creating the equivalent of a weather front in space. Seeing complicated cluster weather like this is the signal of a cluster amalgamation in progress, which may still take a few billion years to complete.

Collisions are not the only action going on in clusters. Many clusters have large, bright galaxies located at their centers, and these galaxies can exert a powerful influence on their surroundings. One such cluster is shown in Figure 7.8a. The Hydra A cluster has a bright, central elliptical galaxy called Hydra A, which contains an active galactic nucleus with a supermassive black hole. Just like the AGNs discussed in Chapter 6, the black hole is responsible for driving tremendous radio jets (fig. 7.8b) that speed outward and slam through the hot cluster gas, pushing it aside and creating large holes about 90,000 light years across (fig. 7.8c). This galaxy is clearly having a devastating effect on its environment, but all of the excitement would be hidden from us if we only observed this cluster in visible light.

Another amazing cluster is the Perseus cluster, located 320 million light years away (fig. 7.9a). The central galaxy, called

Perseus A, also contains a strong AGN powered by a supermassive black hole. The x-ray image shows the fascinating details of how a black hole can effect its surroundings on scales as large as hundreds of thousands of light years. The two large dark holes in the x-ray image are the telltale cavities or bubbles that have been created by the powerful jets of strong high-energy particles spawned by the black hole. Astronomers believe that the ripples of x-ray light that spread outward from the bubbles are evidence that this black hole and its jets actually produce giant sound waves, with a wavelength of 30,000 light years. These waves translate to a B-flat, fifty-seven octaves below middle C on the piano. A piano has only seven octaves, so this sound would be much too low for humans to hear. The sound waves may also carry energy throughout the cluster and continuously heat the gas, keeping it from cooling down and forming more stars. This process may limit the size of the central galaxy and keep it from growing further.

The cluster Abell 2052 is similar to Perseus (fig. 7.10), but the giant cavities have very bright edges, suggesting that the gas here is much more compressed. Not all cavities are the result of ongoing black hole activity, however. The cluster Abell 2597 has cavities (fig. 7.11), but in this case, they are fossils, made about 100 million years ago. These cavities are filled with very hot gas, which makes them much more buoyant than their surroundings. So, like a helium balloon or bubbles in champagne, they rise outward from the center of the cluster. The supermassive black hole at the heart of the central galaxy may eventually become active again and send enough energy outward to create new cavities.

Besides looking for evidence of black hole activity, x-ray images are also a good way to catch galaxies in motion. They may leave behind long trails that act as a footprint of their passage through a cluster, just like the trails left by planes in the sky or boats in the water. The central galaxy in the Centaurus cluster is not standing still. Despite its regular appearance in visible light (fig. 7.12a), the x-ray image (fig. 7.12b) shows that it has a large tail or plume that is twisting around to the left. The plume is approximately 70,000 light years long and contains a mass equal to about 1 billion Suns. It is also several million degrees cooler than the surrounding gas. Astronomers believe this gas is either falling onto the galaxy or being stripped or pushed out of the galaxy as it moves slowly along. If the material is falling onto the galaxy, it may eventually become part of the galaxy and will be available to form new stars, which will allow the galaxy to grow even larger.

The fast-moving central galaxy in Abell 1795 (fig. 7.13) presents even more dramatic behavior. The x-ray image (fig. 7.13c) shows a bright wavy filament 200,000 light years long, made up of dense gas that is about 20,000,000°C cooler than the surrounding gas in the cluster. This "wiggle" trails away from a bright point source at the top, which is an enormous elliptical galaxy (fig. 7.13b). As the galaxy moves through the cluster at a speed of several hundred kilometers per second, the pull of its gravity draws the hot cluster gas around behind it into a giant cosmic wake, like that of a boat in water. This galaxy also has an AGN and the jets are bent backward by the force of its movement through the cluster (fig. 7.13d).

The wide variety and number of x-ray images of clusters are wonderful tools for astronomers, because they can be used to study how galaxies interact with each other and their surroundings. There is plenty of evidence for galaxy wakes, pressure fronts, and shock waves as galaxies plow through the hot gas. Large central galaxies with active supermassive black holes also appear to have a strong interrelationship with the cluster gas, and the explosive behavior of their black holes may be able to regulate the heating and cooling of the gas. But all of this detailed information has been discussed for fairly nearby objects. To know what the early universe was like, we need to look further back in time.

FIGURE 7.1. A crowded neighborhood. One of the densest known clusters of galaxies is the gigantic Coma cluster, roughly 20 million light years across and 370 million light years from Earth, in the constellation Coma Berenices. There are literally thousands of galaxies here, and the visible image (a) shows that they appear to have plenty of empty space between them. The illusion is exposed however in the intensity-colored x-ray image taken with the XMM-Newton satellite (b), where we see that the galaxies are swimming in an immense pool of heated gas with a temperature of about 100,000,000°C. The hot gas is held to the cluster by gravity. Gravity also causes clusters like Coma to continue to grow as they attract still more galaxies. The smaller cloud of gas in the lower right is a neighboring group of galaxies that is being pulled toward Coma by gravity and will eventually become part of the giant cluster.

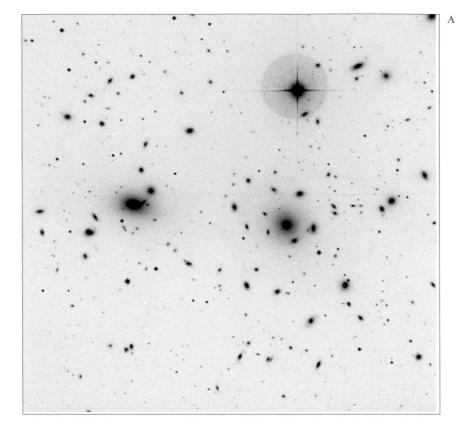

A

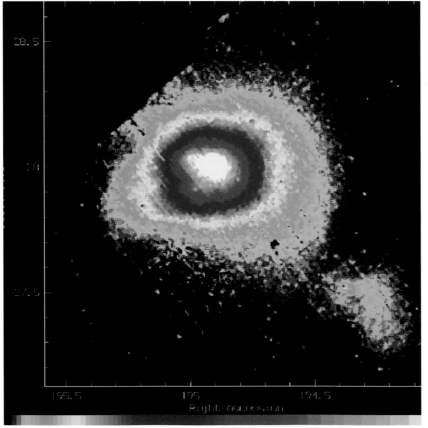

B

FIGURE 7.2 (opposite). Untidy galaxies. This intensity-colored x-ray image taken with Chandra zooms in on the central region of the Coma cluster, about 1.5 million light years in size, where the x-ray light from the hot gas is brightest. The two bright white spots are 10,000 light years across and are dense cool gas clouds with a temperature of only about 10,000,000°C to 20,000,000°C. The cool gas is associated with the two large elliptical galaxies in the visible image. Astronomers believe these giant gas clumps are the remains of material violently ejected from stars similar to the process occurring in NGC 1569 (fig. 5.7). If so, it would have taken about 1 billion years for the stars to eject this much material.

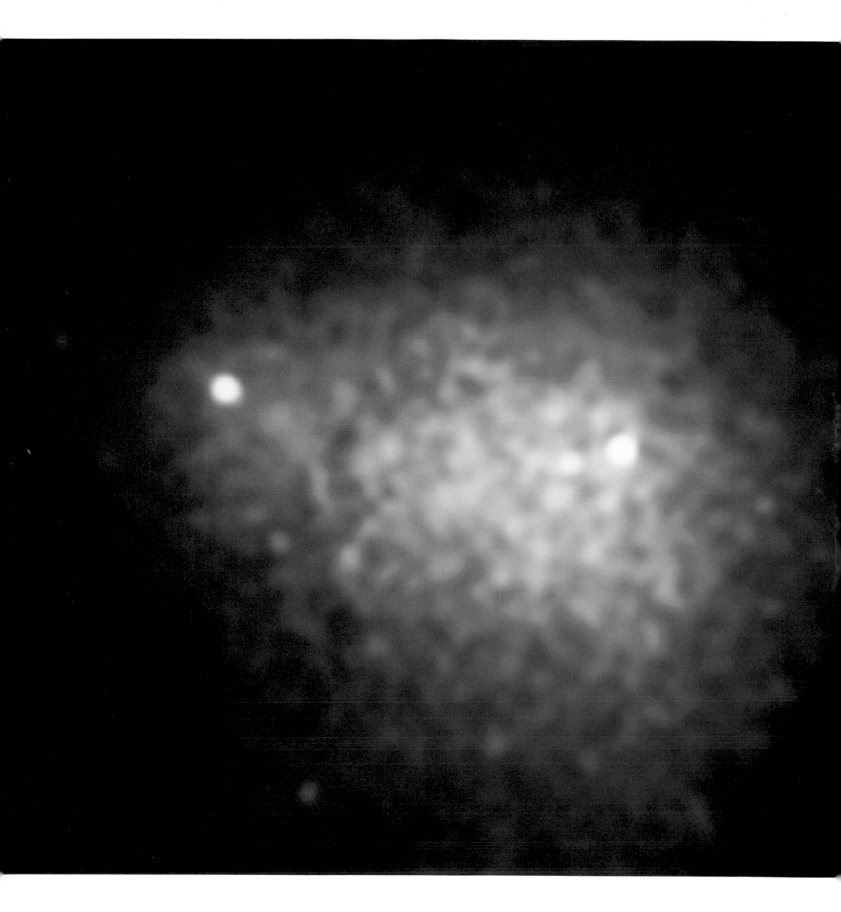

FIGURE 7.3 (opposite). Stephan's Quintet. This small but elegant group of galaxies in the direction of the constellation Pegasus was first discovered in 1877 by the French astronomer Édouard M. Stephan. The group is called Stephan's Quintet, but you'll notice there are six large galaxies rather than five. Stephan originally thought that the five galaxies in the center of the image were all part of the same neighborhood. But while the four yellow galaxies are 280 million light years away, the bluish spiral galaxy is only 40 million light years away, which means that it lines up with the other galaxies only by chance. This cosmic accident threatened to turn the quintet into a mere quartet, but astronomers recently learned that the small rounded galaxy to the left of the image is also 280 million light years away. Thus the quartet is a quintet again.

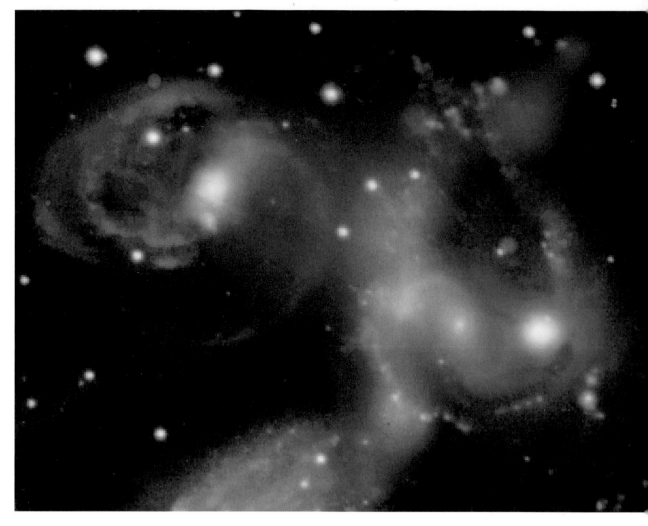

FIGURE 7.4. A cosmic crash. This x-ray image of the Stephan's Quintet placed on top of the visible image shows the dramatic result of a cosmic near-collision. The blue cloud of x-ray light indicates that the gas between the galaxies has been heated to a temperature of 6,000,000°C by the tremendous shocks produced by the galaxies as they slam into it. The main culprit in this crash is the fast-moving spiral galaxy just to the right of the blue cloud, which is speeding through the gas at roughly 2.2 million miles per hour (1,000 kilometers per second). As the galaxies rapidly move through the hot gas, some of the cooler gas in the galaxies will be violently ripped out and some of it will form new stars.

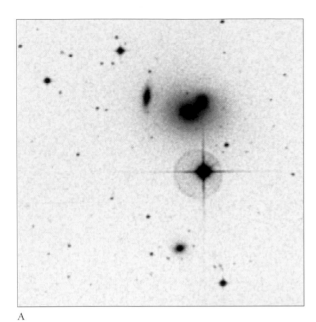

A

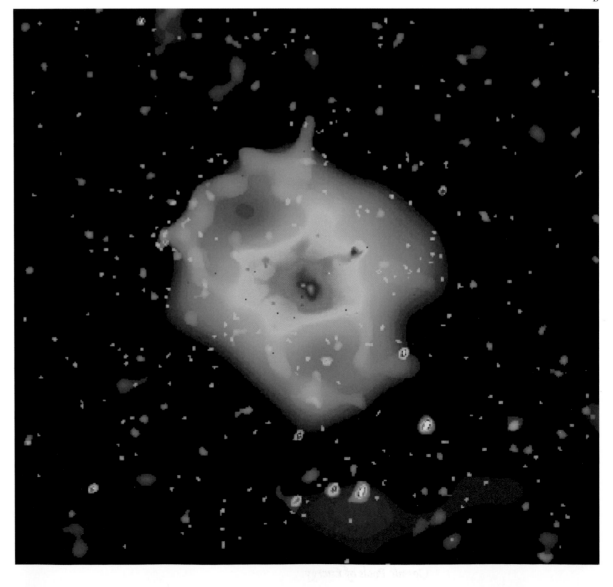

B

FIGURE 7.5. Building blocks. This small group of galaxies known as HCG 62 (a) is located 200 million light years away in the direction of the constellation Virgo. Galaxy groups like this one and Stephan's Quintet are important for understanding the largest structures in the universe, because they are the building blocks for galaxy clusters like Coma. The Chandra intensity-colored image of HCG 62 (b) looks very different from Stephan's Quintet. The gas has an odd shape, with two large holes about 30,000 light years across on either side of the central elliptical galaxy. The holes were probably created by the force of high-energy particles that are produced near the supermassive black hole in the central galaxy.

A

FIGURE 7.6. Dark matter. The cluster of galaxies Abell 2029 is 1 billion light years away in the direction of the constellation Serpens. Like the Coma cluster (fig. 7.1), it contains thousands of galaxies (a). The colossal galaxy at the center of the cluster was probably formed from a union of many smaller galaxies. The intensity-colored x-ray image (b) brightens dramatically toward the center, which tells us that the hot gas between the galaxies is heavily concentrated there. The compact source of x-ray light gives away the presence of large amounts of unseen dark matter condensed toward the center of the cluster—about one hundred trillion Suns' worth. Based partly on observations of clusters like this, astronomers estimate that as much as 90 percent of all the matter in the universe is mysterious dark matter.

B

FIGURE 7.7. Cluster weather. The galaxy cluster Abell 2142 is 6 million light years across but is actually made up of two smaller clusters that are in the midst of crashing together. The arrows in the visible image (a) point to the largest galaxies in these two clusters, while the blob of heated x-ray gas in the intensity-colored x-ray image (b) marks the spot where clusters are colliding and blending together. The gas in the centermost white region in (b) has a temperature of about 50,000,000°C and sits within a larger gas cloud with temperatures of about 70,000,000°C and 100,000,000°C. The cool gas is caused by pressure fronts at the site of the collision, which imply a gigantic weather system, similar to weather systems on Earth.

A

B

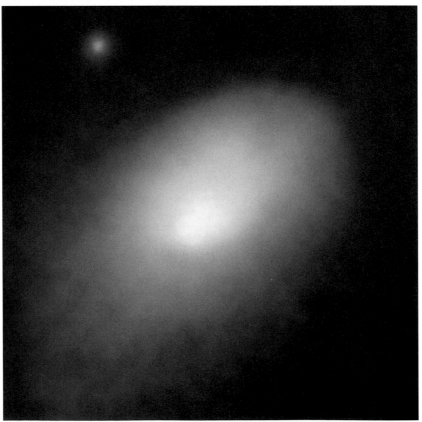

Pages 155—157

FIGURE 7.8. Carving holes. The Hydra A Cluster, 840 million light years away in the constellation Hydra, contains over 100 bright galaxies. It differs from the other clusters discussed so far because its central galaxy has a powerful active galactic nucleus. The visible image of the cluster (a) is dominated by the elliptical galaxy Hydra A, which has an active supermassive black hole at its core. The radio image (b) zooms in on Hydra A and its two colossal radio jets. These jets make a dramatic impact on the cluster gas, as seen in the Chandra intensity-colored x-ray image (c). In the cloud of hot gas in the center of the cluster are two opposing cavities, about 90,000 light years across, carved out by the jets plowing through the gas.

A

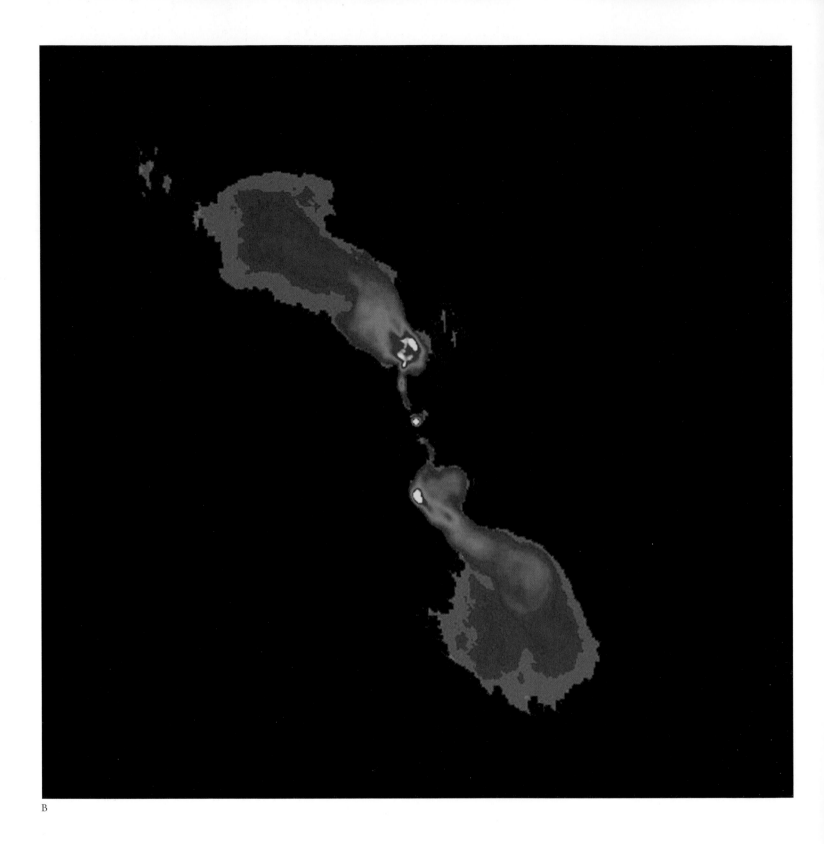

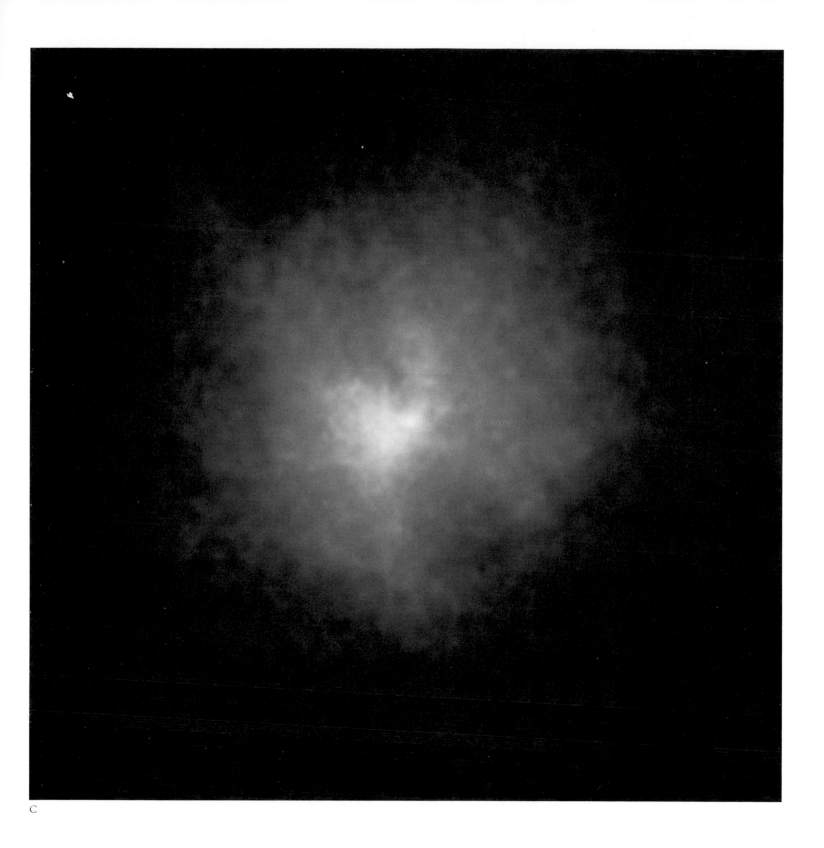

C

FIGURE 7.9. Giant bubbles #1. Located 320 million light years away, the Perseus cluster of galaxies also contains a bright, giant galaxy at its center (a). This galaxy has a supermassive black hole at its core, which apparently produces periodic explosive activity every 10 million years or so. The black hole has a dramatic effect on the cluster, obvious in the energy-colored x-ray image (b), which shows a complicated pattern of x-ray light. The white spot in the center is x-ray light from gas orbiting the black hole. The stretched-out dark patches near the center are shadows cast by the remains of a galaxy that is probably being cannibalized by the central galaxy. The two giant holes in the x-ray gas are bubbles of high-energy particles that are 50,000 light years across. These bubbles have been blown into the surrounding gas by the explosive activity of the black hole, like bubbles blown through a straw underwater. The periodic explosions create ripples with a wavelength of about 30,000 light years. These waves have already traveled hundreds of thousands of light years away from the black hole. Translated into sound, this note would be B-flat, 57 octaves below middle C on the piano—much too low for human hearing, of course!

A

B ➤

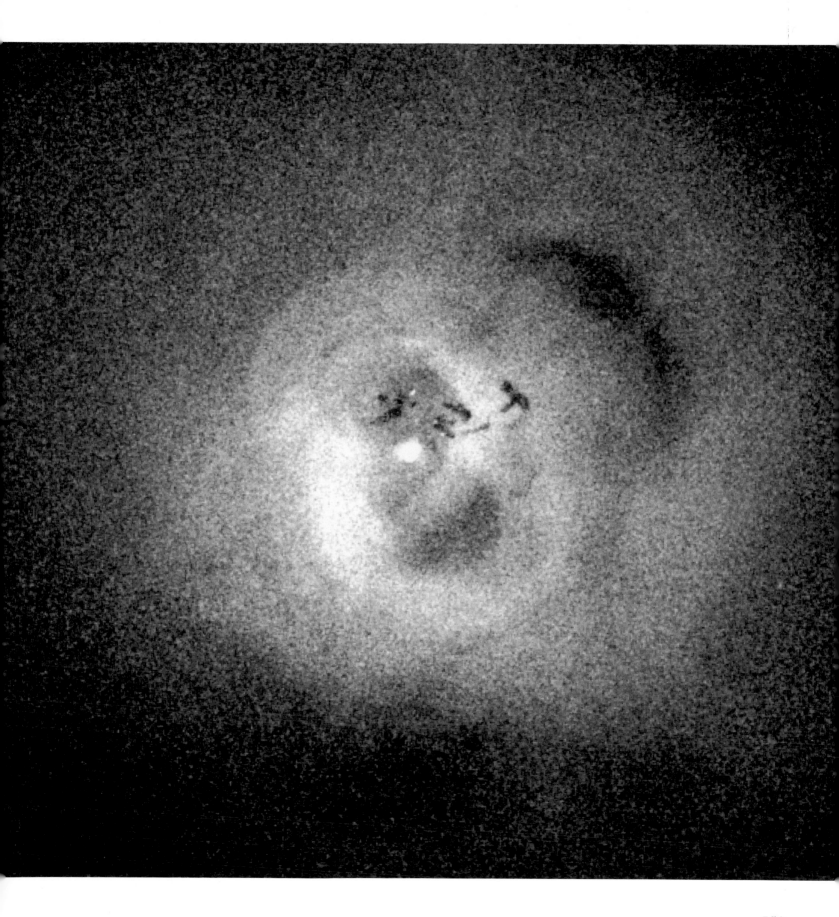

FIGURE 7.10. Giant bubbles #2. The cluster Abell 2052 (a) exhibits behavior that is very similar to Perseus in the sense that the radio source from the black hole in the central galaxy is interacting in a major way with the cluster gas. The intensity-colored x-ray image (b) shows holes in the x-ray gas surrounded by bright dense shells. In this case, the holes are about 60,000 light years across. Immense jets of radio particles expand into the gas, displacing and compressing it, while the hot gas also confines the jets and keeps them from breaking through.

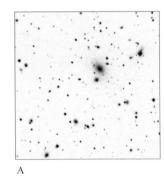

A

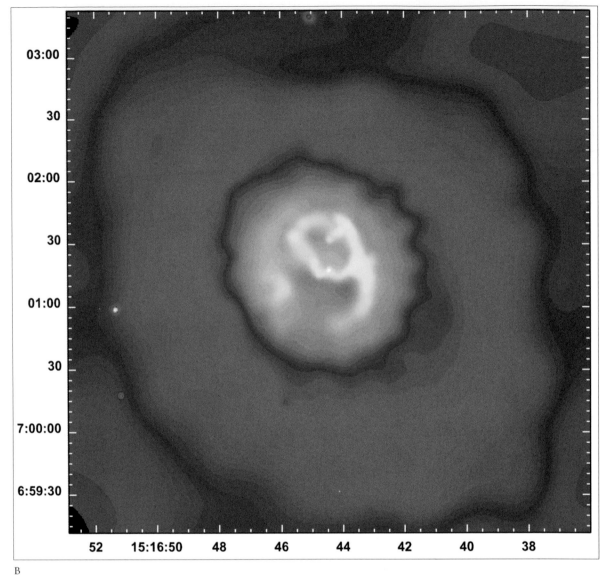

B

A

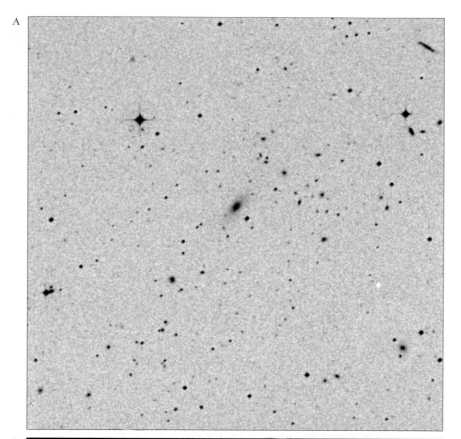

FIGURE 7.11. Ghost holes. The cluster of galaxies Abell 2597 (a) is about 1 billion light years away in the direction of the constellation Aquarius. Zooming in on the central galaxy, the intensity-colored x-ray image (b) shows large cavities that are 60,000 light years across, but now about 100,000 light years away from the center of the cluster, bubbling outward through the hot gas. These cavities are thought to be ghost or fossil cavities left over from an ancient explosion about 100 million years ago from the black hole in the central galaxy. Far from being empty, the fossil holes are filled with high-energy particles and very hot gas. This makes them less dense than the surrounding gas and so they rise outward, like a helium balloon rising in air. Over time, new explosive events in the central galaxy may create new bubbles similar to Perseus and Abell 2052.

B

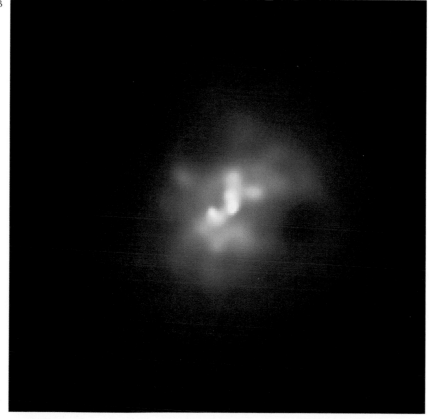

FIGURE 7.12. Cooling gas. The Centaurus cluster is located 170 million light years from Earth in the constellation Centaurus (a). Zooming in on the large central galaxy, the energy-colored x-ray image (b) shows hot gas surrounding the galaxy, as well as a large tail that stretches away from the center. This tail is about 70,000 light years long, with a temperature of about 10,000,000°C, and contains material equivalent to about 1 billion Suns. The gas in the galaxy is much cooler, as indicated by the redder colors. The central galaxy is actually moving slowly through the cluster, and the tail most likely arises from gas cooling down and condensing in the wake of the central galaxy as it moves through the cluster.

A

B ➤

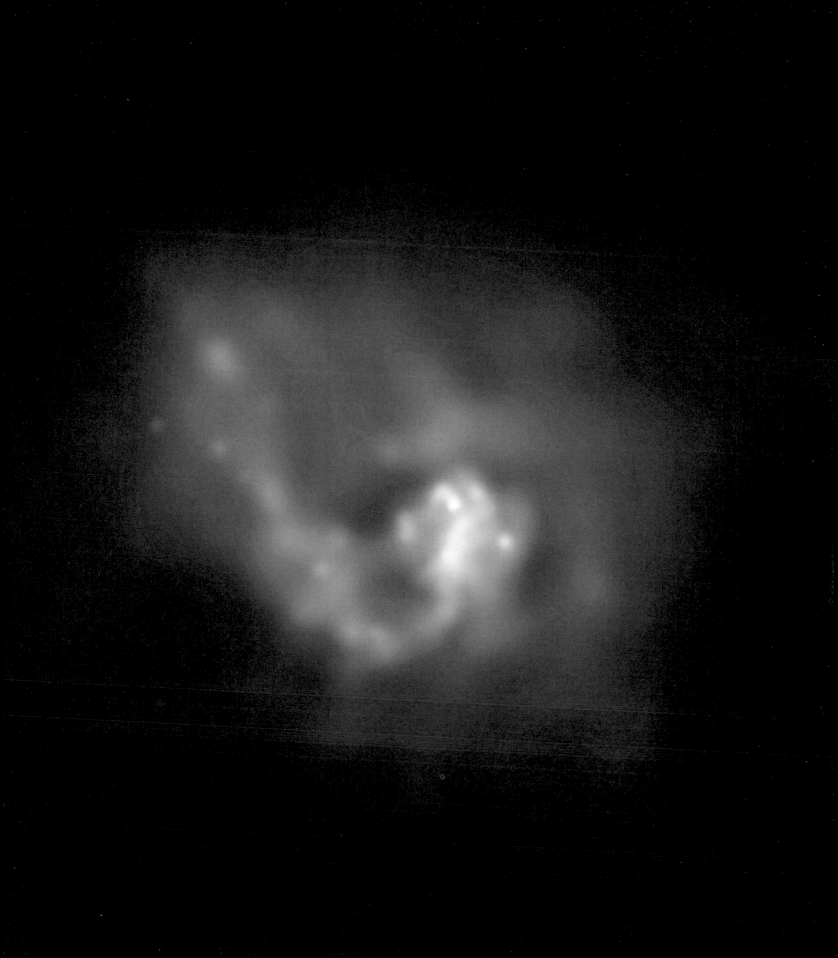

A

B

FIGURE 7.13. A galactic wake. The central galaxy in the cluster Abell 1795 is traveling at a speed of several hundred kilometers per second, cutting a path across the cluster, but the visible image (a) looks inconspicuous. Even zooming in on the galaxy (b), we can't tell that it's moving. However, the intensity-colored x-ray image (c) clearly shows the path of the galaxy as a bright "wiggle" of x-ray light. The size of the galaxy corresponds to the topmost white part of the x-ray wiggle. The wiggle is 200,000 light years long and is denser and cooler than the rest of the gas in the cluster. When the x-ray image is combined with the outline of the radio emission from the galaxy (d), we see that the jets are being bent backward by the hot gas like branches bending in the wind. As the galaxy speeds across the cluster though the hot gas, gravity pulls the gas around behind it, and the gas is left in a dense, cool path in its wake.

C

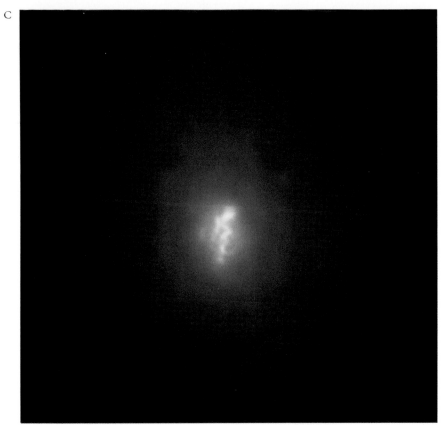

D

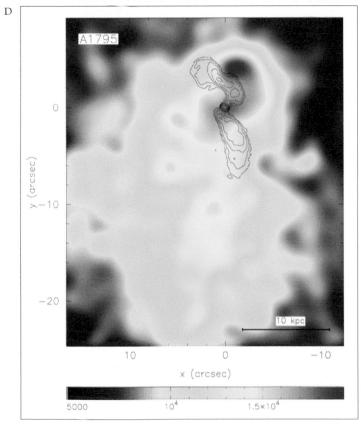

8

The Ancient Past

SO FAR, the celestial objects discussed in this book have been mostly limited to those that are less than 1 billion light years away. Scientists believe that the universe is about 13.7 billion years old, so these objects represent stars, galaxies, and clusters of galaxies as they appear to us today, spanning only about the last 7 percent of the universe's current age. Nearby objects are the easiest for astronomers to study because our telescopes can take high-quality pictures of them, but to learn where these objects came from, we must look much farther back in time. In human terms, examining only the nearby universe is like examining only the last 5 years of a 70-year-old person's life without asking any questions and trying to determine who that person's parents were, where the person was born, or what games they played as a child. But as we look deeper into space and farther back in time, most normal galaxies become too faint to see in x-rays with current telescopes. So, we must search for specific signposts, special galaxies that are extremely powerful or that are doing something extremely unusual and are able to transmit their energy across great distances.

As we saw in Chapter 7, clusters of galaxies are enormous, so we can see them at very large distances. In order to know how clusters were formed and how they grew from smaller groups of galaxies coming together, we have to search for dynamic youthful clusters. Something dramatic is indeed happening in the galaxy cluster IE 0657-56, located 4 billion light years away. The x-ray image in figure 8.1 shows a remarkable V-shaped feature that marks the location of a giant shock wave where gas is being ripped from a small group of galaxies that is plowing through the cluster like a bullet at a speed of about 6 million miles per hour (2,700 kilometers per second). This group may be all that's left of the center of an even larger group of galaxies that collided with IE 0657-56 in the past. Watching the warm gas from the galaxies being heated and stripped away by the very hot gas in the cluster is a wonderful illustration of the way in which small galaxy groups and small clusters can come together to form large clusters. The unseen force of gravity is driving this scene.

Active galactic nuclei (AGNs) are also visible at great distances. The most powerful of these are quasars (fig. 6.6a), which can be brighter than thousands of typical nearby AGNs. Astronomers generally believe that a quasar is a young AGN, at a time in its life when its galaxy is forming and the supermassive black hole is feeding prodigiously and hence shining much more brightly than similar black holes do at a later stage. The quasar shown in figure 8.2 is called PKS 0637-752 and is one of the most powerful objects in the known universe. This quasar possesses a mighty x-ray jet that is 300,000 light years across, but the source of energy for this jet remains a mystery. Because it is so far away from us, we see PKS 0637-752 as it was 6 billion years ago.

The quasar PKS 1127-145 (fig. 8.3a), another powerful source of x-rays, is located about 10 billion light years from Earth. The colossal one-sided x-ray jet in this galaxy covers a distance of about one million light years, which is about ten times the diameter of the Milky Way galaxy. The jet is a vibrant signpost of a supermassive black hole and is produced by explosive activity somehow connected to the gas that is swirling around the central black hole. Over time, the explosions near the black hole expel giant blobs of material outward along the direction of the jet. The immense size of the jet and its lumpy appearance tell us that although this explosive activity has continued for at least a million years, the black hole has taken an occasional break.

Because supermassive black holes in quasars are responsible for large amounts of x-ray light, x-ray images can be used to uncover them. The visible image of the region of sky containing PKS 1127-145 (fig. 8.3b) does not directly reveal the presence of the quasar. The difference between the visible and x-ray images is extreme and shows how x-rays might allow astronomers to find quasars at vast distances. X-ray astronomers are lucky, because both the violent effects of galaxy-wide star formation (Chapter 5) and active black holes in AGNs (Chapter 6) are intense sources of x-rays. These beacons of x-ray light provide direct clues about galaxy formation and the advent of the first black holes.

When it comes to examining the ancient universe on grander scales, visible and x-ray pictures do not display the same information. Figure 8.4a shows a region of the sky known as the Hubble Deep Field, which is humankind's deepest gaze and most detailed visible picture of our universe so far. Like a time machine, this image shows some galaxies as they looked 10 billion years ago. Only twelve of the galaxies are also x-ray sources (fig. 8.4b). Although some of these contain supermassive black holes, the rest are relatively normal galaxies, where the x-rays come from stars, x-ray binary systems, and supernova remnants. This x-ray image allows astronomers to witness violent stellar processes in a handful of normal galaxies from as much as several billion years ago, which gives us an idea of what our own galaxy may have been like when it was much younger.

Astronomers have looked even farther back in time and have recently captured a remarkable baby picture of the universe. Figure 8.5 is an image of the universe in microwave light taken with NASA's Wilkinson Microwave Anisotropy Probe (WMAP). This image is the earliest snapshot of the universe ever taken and it shows how the universe looked when it was only 380,000 years old, which is just 0.003 percent of its current age. The range of colors symbolize tiny temperature changes in the light that represent the places where matter first began to collect and would eventually grow to form galaxies and clusters of galaxies.

Although our x-ray satellites cannot yet see back as far as WMAP, it is possible to use high-quality x-ray images to learn about the evolution of galaxies. Two of the most sensitive x-ray pictures recorded so far are shown in figure 8.6. The points of x-ray light in these images represent some of the farthest and youngest objects in the universe. Many are related to supermassive black holes in the centers of galaxies, but there are also normal galaxies, starburst galaxies, and groups and clusters of galaxies. Figure 8.6a contains more than 500 individual sources and takes us all the way back to the time when the universe was only 1 billion years old (7% of its current age). The images suggest that supermassive black holes in the early universe are rarer than expected, meaning it takes more time for them to form than astronomers originally believed.

The image in figure 8.7 is a visible image of another region of the sky containing one of the most distant quasars known, about 13 billion light years from Earth. In this image the quasar is extremely faint—seen here as a small red dot. It's much easier to see the quasar in the x-ray image, because the x-rays can pass through the galaxy in which the quasar is located and escape better than visible light can. The x-rays from this quasar were created when the universe was only about one billion years old, so this quasar is very young.

Some distant quasars are not seen at all in visible light, which means that they are buried in large amounts of gas and dust, which may arise from the formation of the galaxy that contains the quasar. Figure 8.8 shows one such object, which gives off x-rays and infrared light but no visible light. The x-ray and infrared light together provide a powerful way to search for black holes far back in time that are buried deeply inside a dusty galaxy that is rapidly forming stars. These images may represent a young galaxy just being born.

Current space-based telescopes have already given us tantalizing clues about how galaxies and clusters of galaxies form and grow. As scientists continue to obtain new images of deep space with even longer exposures, we are certain to learn much more about our universe's ancient past. ✳

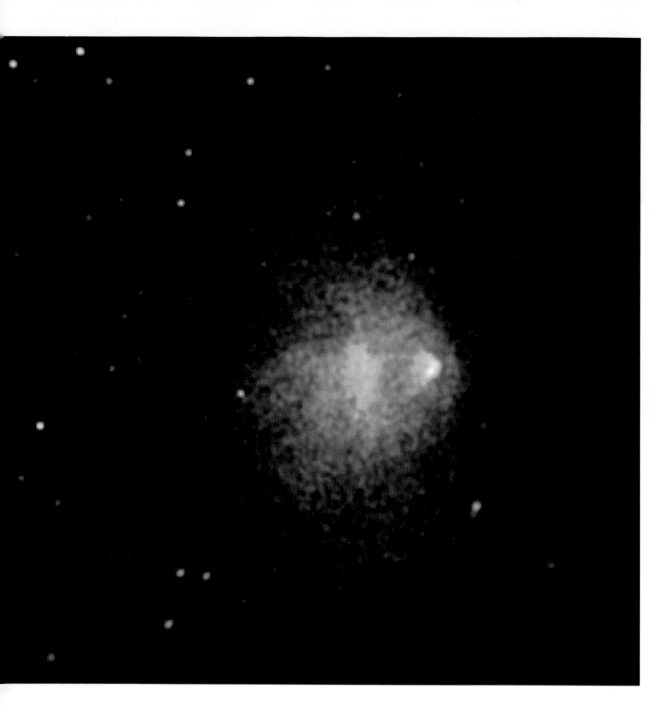

FIGURE 8.1. Speeding bullet. This intensity-colored Chandra x-ray image is a picture of the galaxy cluster called 1E 0657-56, located 4 billion light years away in the constellation Carina. Astronomers believe that the bow-shaped white area that looks like the wake of a speedboat plowing through water is a shock wave formed by a small group of galaxies cutting through the cluster at a speed of 6 million miles per hour (about 2,700 kilometers per second). These speeding galaxies would have passed through the center of the cluster about 150 million years ago. The galaxies carry warm gas with them with a temperature of 70,000,000°C. As this warm gas slams though the hotter, 100-million-degree gas in the cluster, it is blown backward like leaves from a tree in a wind storm. The fast galaxies will eventually be slowed down by gravity and will become indistinguishable from the rest of the cluster.

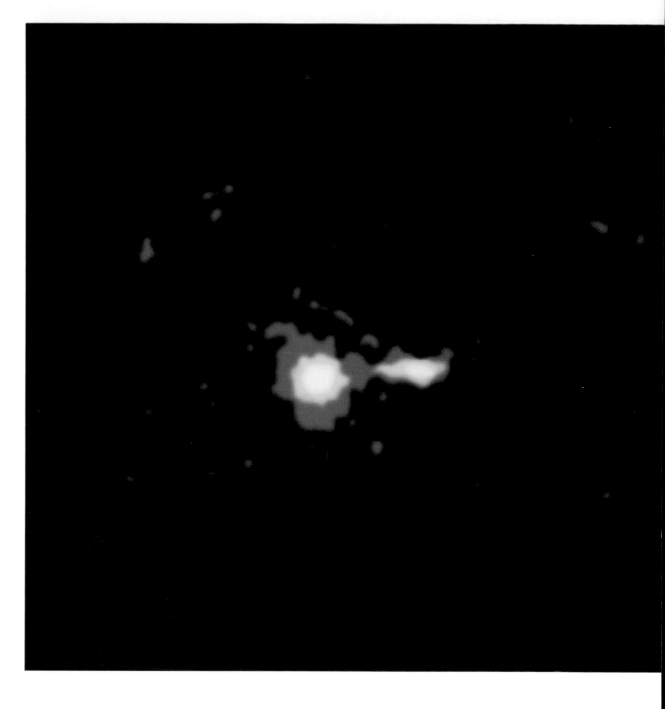

FIGURE 8.2. Powerful quasar. Intensity-colored x-ray picture of the quasar PKS 0637-752 located 6 billion light years away in the constellation Mensa. We are seeing this quasar as it was 6 billion years ago. The amount of energy pouring from the supermassive black hole at the heart of this quasar is equal to that of 10 trillion Suns.

FIGURE 8.3. X-ray beacon. This is the region of sky that contains the quasar PKS 1127-145, which is 10 billion light years away in the constellation Crater. The Chandra intensity-colored x-ray image overlaid on the visible image (a) shows which galaxy is the brilliant quasar—the x-rays serve as a beacon to let us know where the black hole hides. The visible image taken with the Hubble Space Telescope (b) shows the galaxies. The jet of x-rays is huge—about 1 million light years long. The jet is probably due to high-speed, high-energy electrons that are shot away from the black hole and are colliding with low-energy photons that permeate the space between galaxies. The length of the jet tells us that this explosive activity has been going on for a very long time, but its lumpy appearance suggests that the black hole has taken occasional breaks.

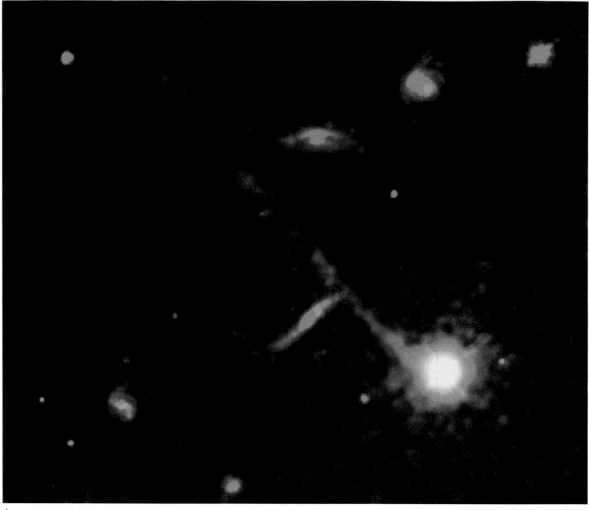

A

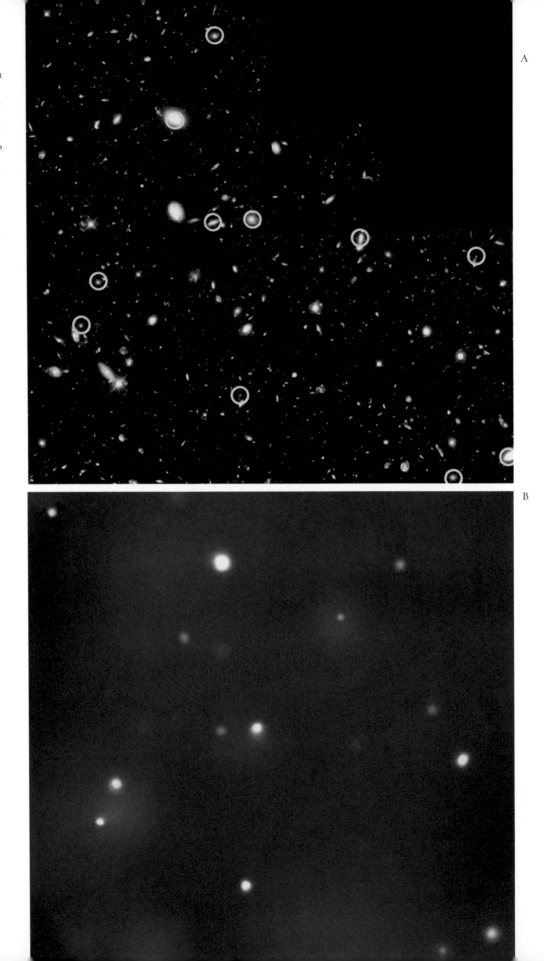

FIGURE 8.4. X-ray sky. Hubble Space Telescope (a) and Chandra (b) image of the night sky. The yellow circles indicate which visible sources are also x-ray sources. This patch of sky falls in a region called the northern Hubble Deep Field, which is the best-studied piece of sky at all wavelengths of light. The x-ray image is energy-colored, with red objects being cooler than blue objects. This image lets us look back a few billion years into the past. Some of the x-ray objects are normal galaxies like our Milky Way and others are supermassive black holes in active galaxies. This x-ray image is a small piece of the northern Chandra Deep Field (fig. 8.6a).

FIGURE 8.5. Baby universe. Image of the baby universe in microwave light obtained by NASA's Wilkinson Microwave Anisotropy Probe, or WMAP. This is the first detailed full-sky mapping of the earliest (oldest) light in the universe. The light in this image is from only 380,000 years after the Big Bang. This map is a temperature map that tracks tiny changes in temperature in the smooth bath of microwave light left over from the Big Bang at a time when the temperature of the universe finally became low enough for atoms to form. The warmer spots are red and the cooler spots are blue. The detailed structures are the seeds of the cosmic structure in the universe we see today, such as galaxies and clusters. The oval shape of the image is a simple way to project the whole sky into a two-dimensional image, just like projecting a map of the Earth's surface.

The Ancient Past

FIGURE 8.6. Long looks. These energy-colored Chandra images of the sky are two of the longest exposures ever taken with an x-ray telescope. Some of the faintest sources here create a photon of x-ray light only once every few days. Figure 8.6a is known as the northern Chandra Deep Field. It is located in the constellation Ursa Major in the Northern Hemisphere, and the image was exposed for two million seconds (23 days). The area of the sky is three-fifths the size of the full Moon (see fig. 4.2). This image takes us all the way back to the time when the universe was only about 1 billion years old. The southern Chandra Deep Field (b) is located in the Southern Hemisphere constellation Fornax. This image was exposed for one million seconds (11 days) and is about half the size of the northern Deep Field. Many of the x-ray sources here are supermassive black holes at the centers of galaxies.

A

B ➤

FIGURE 8.7. Most distant. The quasar SDSS 1030+0524 is one of the most distant. Located 13 billion light years away in the constellation Sextans, this quasar probably contains a black hole with a mass of one to ten billion solar masses. The visible image (a) shows a faint red fuzzy point at the location of SDSS 1030+0524. The quasar shows up better in the x-ray image (b). Since the quasar is so far away, it is also a very young quasar. The x-rays from this quasar were made when the universe was about 1 billion years old.

A

B ➤

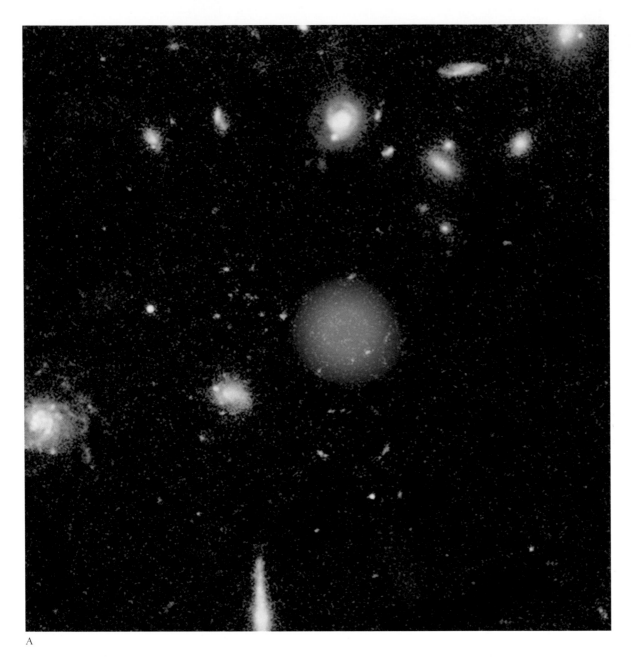

A

FIGURE 8.8. Galaxy formation. This set of combined pictures from Chandra and Hubble (a) and NASA's new Spitzer Space Telescope (b), which takes pictures of infrared light, shows how to find a hidden black hole that is buried deep in a dusty galaxy. These images cover a small portion of the sky from the southern Chandra Deep Field (fig. 8.6b). In the first image, yellow and white represent visible light and blue represents x-ray light. The second image is the infrared image. Infrared light is made by dust that is heated when it absorbs visible light. Comparing the two images, most of the galaxies in the visible image are also seen in infrared light, but the visible image shows no galaxy at the position of the source of the x-rays. On the other hand, there is a bright infrared source that matches the location of the x-ray source. This suggests a giant black hole that is completely buried in a young, dusty galaxy.

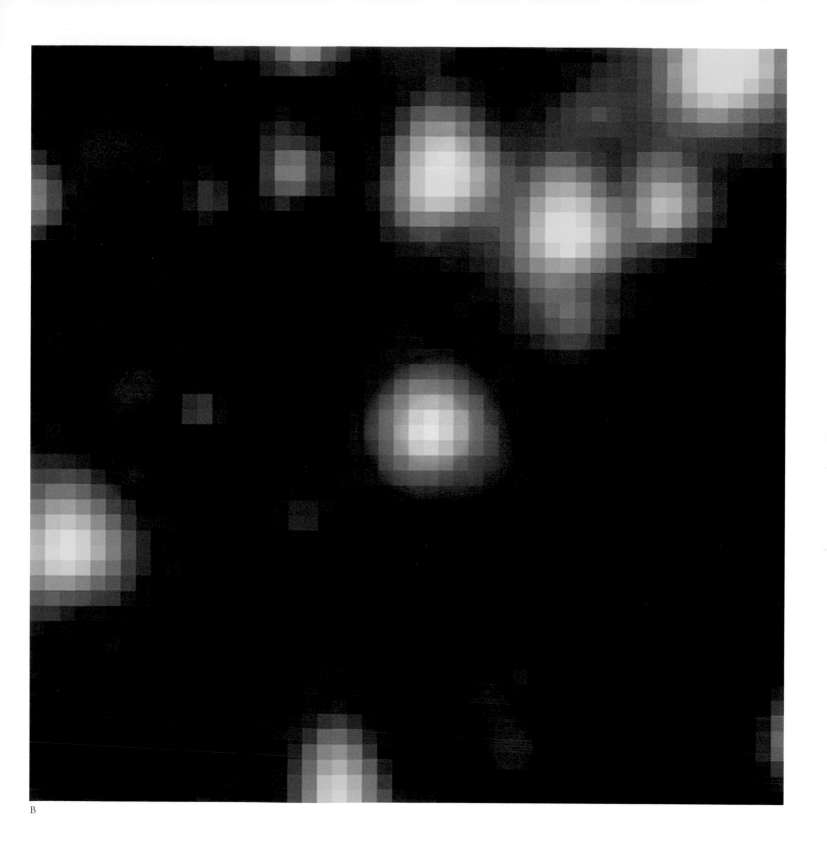

B

The Journey Continues

9

DURING THE past four decades, astronomers have learned a great deal about the previously invisible regions of our cosmos. Just think about how much excitement we would be missing without our x-ray telescopes—our x-ray "eyes" in space. We could not peer into the hearts of galaxies to witness their powerful black holes in action. We would never see the graceful wakes left by galaxies as they speed through immense pools of heated gas in giant clusters. Nor could we peek into protected stellar nurseries to watch the birth of new stars. This expanded knowledge has revolutionized our concept of ourselves, our galaxy, and our universe, but there is still much more to learn. Forty years from now, we shall examine today's x-ray pictures and be astonished at how far we've come, just as today's optical astronomers look at vintage photographs from a century ago and are amazed at how much their telescopic vision has improved.

X-ray astronomers are currently hard at work creating a new generation of better and more powerful x-ray telescopes. The first among this new generation is a small but advanced telescope called Astro-E2 (fig. 9.1), scheduled to launch from Japan in 2005. This telescope was conceived and built by a dedicated team of American and Japanese scientists. Astro-E2 was named for its sister mission, Astro-E, which was lost in an accident soon after it was launched on February 10, 2000 (fig. 9.2). The rocket carrying Astro-E developed a control system failure, and the satellite never reached its intended orbit. Although the

loss of Astro-E was terribly unfortunate for the scientists working on the mission, they would not allow themselves to give in to
defeat. After the accident, they quickly got to work planning and
building a replacement telescope. Astro-E2 was built in record
time and the scientists are now hoping to achieve the mission's
original promise.

Astro-E2 is amazingly powerful; it will have the ability to
split the spectrum of x-ray light into extremely fine slices. This
will allow astronomers to make detailed measurements of the
energy, or wavelength, of an individual x-ray photon. By obtain-
ing such precise measurements, it will be possible to determine
the physical properties of certain cosmic sources with greater
accuracy than ever before. Scientists will measure exactly how
temperatures change in hot gas across nearby clusters of galaxies,
the relative amounts of chemical elements produced in stars and
spread out in supernovae, and the speeds of material orbiting
near the edge of a black hole.

X-ray missions that are even more powerful than Astro-
E2 are on the drawing board for launch by the end of the next
decade. Two of these, named Constellation-X (fig. 9.3a) and the
X-ray Evolving Universe Spectrometer, or XEUS (fig. 9.3b), will
combine the best characteristics of current x-ray missions such
as Chandra and XMM-Newton. These telescopes will allow
x-ray astronomers to see farther into space than ever before.
Constellation-X is a NASA mission that is designed to take pictures that look much like those from XMM-Newton (see, e.g.,
fig. 7.1b), while collecting hundreds of times more x-rays than
Chandra. It will also be able to split x-ray light into even finer
slices than Astro-E2 and will collect x-rays that are up to six
times more powerful than those captured by Chandra.

XEUS is being designed by the European Space Agency
to make high-resolution pictures as good as Chandra while capturing one hundred times more x-rays (similar to Constellation-
X). The pictures from XEUS will allow us to peer farther back
in time than the pictures from Chandra, and astronomers will

be able to use these pictures to see extremely fine details in complicated regions of space such as the center of our galaxy (Chapter 4) and the centers of clusters of galaxies (Chapter 7). The leap in our ability to detect subtle details provided by both Constellation-X and XEUS will be similar to the improvement that optical and infrared astronomers will experience with the James Webb space telescope, which is NASA's planned replacement for the Hubble Space Telescope.

Since Constellation-X and XEUS are in the planning stages, their concepts may change and evolve as they come closer to completion. But regardless of their final forms, their goals will undoubtedly remain the same—to collect as many x-ray photons as possible, taking the best pictures possible, and splitting x-ray light into very fine slices, while extending our reach toward even more powerful x-rays. Together, XEUS and Constellation-X will allow astronomers to look back to times before galaxies were born, study the formation of galaxies and clusters of galaxies, determine how black holes evolve, and measure the masses of black holes and the extreme gravitational fields surrounding them.

Beyond these two missions, scientists are already planning ideas for 25 to 50 years from now. An even more ambitious mission will be able to take a picture of the edge, or event horizon, of a black hole. Such pictures require a resolution that is as much as a million times better than Chandra, making the mission a technical challenge. But these pictures will let us take the ultimate virtual journey through space to visit a black hole. It will finally be possible to see the dramatic distortion of space predicted by Albert Einstein, the bending of light by the extreme gravity of a black hole (fig. 9.4).

With these missions to come, the outlook of x-ray astronomy is bright. Adding this high-energy window to the many other ways in which we are now seeing our universe has made astronomy a truly comprehensive reservoir of information. Although our questions about the universe have existed for thousands of years, a complete view of the universe has only been made possible recently, within a mere human lifespan. We are in the middle of a scientific Renaissance, with our minds coming ever closer to understanding the mysteries of our universe's past and the course of our future.

FIGURE 9.1. Astro-E2. This is an artist's drawing of the Astro-E2 satellite as it will look when it is orbiting Earth. Astro-E2 will be the next x-ray telescope to fly in space, and it is scheduled to launch in 2005 from Japan.

FIGURE 9.2. First try. The launch of the Astro-E satellite on February 10, 2000, in Japan, lifting off from the Kagoshima Space Center on the Japanese island of Kyushu. Unfortunately, the control system of the rocket failed during the first portion of its flight, and the mission aborted shortly after takeoff. The rocket was not able to carry the satellite to a high enough altitude; it ended up, unusable, in an orbit that was much too low. Scientists have built Astro-E2 to take its place.

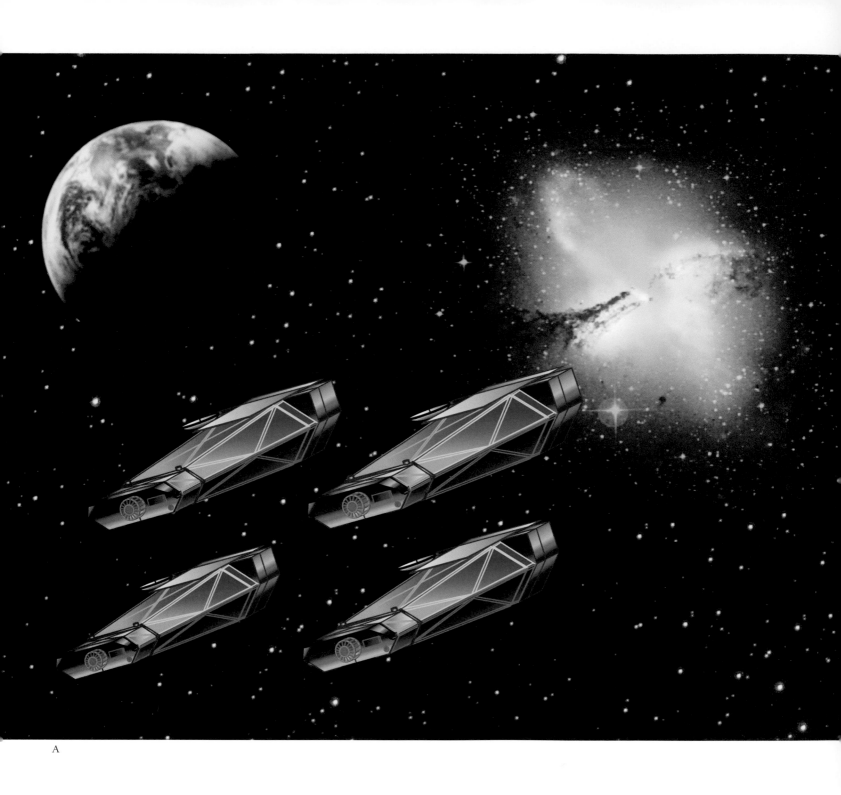

A

FIGURE 9.3. Future prospects. Two ideas for powerful future x-ray missions are Constellation-X (a) and XEUS (X-ray Evolving Universe Spectrometer) (b). Constellation-X is designed to collect 100 times more x-ray photons than Chandra and to divide x-ray light into pieces that are three to four times smaller than those divided by Astro-E2. XEUS will also collect many more photons than Chandra, while creating very high quality images. Both of these missions will advance the field of x-ray astronomy by allowing astronomers to look farther back in time to watch the universe being formed and to capture light from near the event horizons of black holes.

B

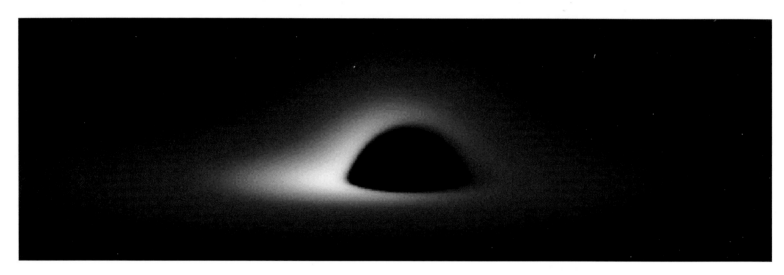

FIGURE 9.4. Portrait of a black hole. One of the ultimate goals of x-ray astronomy is to take a picture of the edge of a black hole so that we can determine if it looks the way Einstein's equations predict. The extreme gravity of a black hole should bend light so strongly that it causes us to see the other side of the material falling in—in this case we would view the other side of the accretion disk bent up around the black hole, distorting its shape. Since supermassive black holes in other galaxies are only about the size of our solar system, it will take excellent vision to obtain such a picture—about one million times better than what is currently available with a telescope like Chandra.

Index

Page numbers for illustrations are in *italics*.

CREDITS

Acronyms used in the listing below are as follows:

AA: Astronomy and Astrophysics

AAS: American Astronomical Society

AFRL: Air Force Research Laboratory

AJ: Astronomical Journal

ApJ: Astrophysical Journal

ApJS: Astrophysical Journal Supplement

ASU: Arizona State University

AUI: Associated Universities, Inc.

AURA: Association of Universities for Research in Astronomy

CfA: Harvard-Smithsonian Center for Astrophysics

CIW: Carnegie Institution of Washington

CXC: Chandra X-ray Center

DSS: Digitized Sky Survey

DTM: Department of Terrestrial Magnetism

ESA: European Space Agency

GOODS: The Great Observatories Origins Deep Survey

GSFC: Goddard Space Flight Center

HST: Hubble Space Telescope

INAF: National Institute for Astrophysics

IOA: Institute of Astronomy, Cambridge

ISAS: Institute of Space and Astronautical Science

JAXA: Japan Aerospace Exploration Agency

JHU: Johns Hopkins University

MIT: Massachusetts Institute of Technology

MPE: Max Planck Institute for Extraterrestrial Physics

MSX: Midcourse Space Experiment

NAO: National Astronomical Observatory

NASA: National Aeronautics and Space Administration

NGST: Next Generation Space Telescope

NMSU: New Mexico State University

NOAO: National Optical Astronomy Observatory

NRAO: National Radio Astronomy Observatory

NRL: Naval Research Laboratory

NSF: National Science Foundation

NWU: Northwestern University

PSU: Pennsylvania State University

REU: Research Experiences for Undergraduates

ROSAT: Roentgen Satellite

SAO: Smithsonian Astrophysical Observatory

SOHO: Solar and Heliospheric Observatory

ST-ECF: Space Telescope–European Coordinating Facility

STSCI: Space Telescope Science Institute

UCI: University of California, Irvine

UCSB: University of California, Santa Barbara

UCSD: University of California, San Diego

UIT: Ultraviolet Imaging Telescope

UIUC: University of Illinois at Urbana–Champaign

UMBC: University of Maryland, Baltimore County

UMd: University of Maryland

VLA: Very Large Array

WIYN: Wisconsin Indiana Yale NOAO Consortium

WMAP: Wilkinson Microwave Anisotropy Probe

Fig. 1.1: NASA

Fig. 1.2: NASA (original), redrawn by JHUP

Fig. 2.1–2.2: provided by author

Fig. 2.3: (a) R. Giacconi and H. Gursky, Space Science Reviews 4 (1965): 151, fig. 3 (Kluwer Academic Publishers, with kind permission of Springer Science and Business Media); (b) MPE/ROSAT

Fig. 2.4: (a) NASA/GSFC/D. Batchelor; (b, c) NASA/GSFC/JHU/K. Weaver

Fig. 2.5: (a) CXC/NGST; (b) Courtesy of ESA

Fig. 2.6–2.8: NASA/CXC/JHU/K. Weaver

Fig. 3.1: (a) Yohkoh Science Team; (b) SOHO(ESA and NASA)

Fig. 3.2: NASA/J. Hester and P. Scowen (ASU)

Fig. 3.3: (a) NASA/C. O'Dell (Vanderbilt University) and S. Wong (Rice University); (b) NASA/PSU

Fig. 3.4: NASA/GSFC/M. Corcoran et al.

Fig. 3.5: (a) T. Rector/University of Alaska Anchorage, WIYN, and NOAO/AURA/NSF; (b) NASA/PSU/L. Townsley et al.

Fig. 3.6: (a) NASA/HST/J. Morse/K. Davidson; (b) NASA/CXC/SAO

Fig. 3.7: (a) NASA/UIUC/Y. Chu et al.; (b) x-ray: NASA/UIUC/Y. Chu et al., visible: NASA/HST

Fig. 3.8: (a, c, d) NASA/CXC/SAO; (b) NASA/CXC/Rutgers U./J. Hughes et al.

Fig. 3.9: NASA/GSFC/U. Hwang et al.

Fig. 3.10: (a) VLA/NRAO; (b) W. M. Keck Observatory; (c) Palomar Observatory; (d) NASA/CXC/SAO

Fig. 3.11: x-ray: NASA/CXC/ASU/J. Hester et al., visible: NASA/HST/ASU/J. Hester et al., radio: VLA/NRAO

Fig. 3.12: NASA/CXC/U. Mass/F. Lu et al.

Fig. 3.13: NASA/SAO/CXC

Fig. 4.1: NASA and G. Bacon (STScI)

Fig. 4.2: REU program, N. Sharp/NOAO/AURA/NSF

Fig. 4.3: NASA/CXC/SAO

Fig. 4.4: CXC/S. Lee

Fig. 4.5: Celestial Images

Fig. 4.6: NASA/U. Mass/D. Wang et al.

Fig. 4.7: (a) VLA/NRL/N. Kassim; (b) MSX/AFRL/S. Price; (c) NASA/U. Mass/D. Wang et al.; (d) x-ray: NASA/U. Mass/D. Wang et al., radio: VLA/NRL/N. Kassim, infrared: MSX/AFRL/S. Price

Fig. 4.8: NASA/MIT/F. Baganoff et al.

Fig. 4.9: NRAO/NWU/F. Yusef-Zadeh

Fig. 4.10: CXC/M. Weiss

Fig. 4.11: NASA/CXC/U. Amsterdam/S. Migliari et al.

Fig. 4.12: NASA/CXC

Fig. 5.1: (a) Anglo-Australian Observatory/David Malin Images; (b, c) NASA/CXC/U. Leicester/U. London/R. Soria and K. Wu

Fig. 5.2: (a) T. Rector and M. Ramirez/NOAO/AURA/NSF; (b) NASA/CXC/UMd/A. Wilson et al.; (c) NASA/AURA/STScI/Hubble Heritage Team; (d) NASA/CXC/UMd/A. Wilson et al.

Fig. 5.3: (a) F. Schweizer (CIW/DTM); (b) NASA/HST/STScI/B. Whitmore et al.; (c) NASA/SAO/G. Fabbiano et al.

Fig. 5.4: (a) Palomar Observatory DSS*; (b) NASA/CXC/A. Zezas et al.

Fig. 5.5: NASA/CXC/STScI/U. North Carolina/G. Cecil

Fig. 5.6: (a) NOAO/AURA/NSF; (b) NASA/CXC/U. Mass/D. Wang et al.; (c) x-ray: NASA/CXC/U. Mass/D. Wang et al., visible: NASA/HST/D. Wang et al., ultraviolet: NASA/GSFC/UIT, black-and-white visible: NOAO/AURA/NSF

Fig. 5.7: (a) C. Martin, H. Kobulnicky, and T. Heckman, ApJ 574 (2002): 663 (reproduced by permission of the AAS); (b) NASA/CXC/UCSB/C. Martin et al.

Fig. 5.8: (a) Subaru Telescope, NAO Japan; (b) NASA/SAO/ G. Fabbiano et al.

Fig. 5.9: NASA/SAO/CXC

Fig. 5.10: (a) T. Boroson/NOAO/ AURA/NSF; (b) NASA/SAO/CXC; (c) NASA/CXC/JHU/K. Weaver

Fig. 6.1: CXC/A. Hobart

Fig. 6.2: NASA/R. van der Marel (STScI)

Fig. 6.3: NASA/A. Kamajian

Fig. 6.4: (a) STScI/DSS/UK Schmidt**; (b) NRAO/VLA/J. Condon et al.; (c) NASA/CXC/SAO/M. Karovska et al.; (d) all of the above

Fig. 6.5: NASA/STScI/E. Schreier

Fig. 6.6: (a) NOAO/AURA/NSF; (b) NASA/CXC/SAO/H. Marshall et al.

Fig. 6.7: (a) NOAO/AURA/NSF; (b) NASA/ STScI/UMBC/E. Perlman et al.; (c) J. Biretta, F. Zhou, and F. Owen, ApJ 447 (1995):582 (reproduced by permission of the AAS); (d) NASA/CXC/MIT/H. Marshall et al.

Fig. 6.8: (a) ST-ECF, Munich/R. Fosbury and R. Hook; (b) VLA observations by J. Conway (NRAO/VLA); and P. Blanco (UCSD); (c) A. Wilson, A. Young, and P. Shopbell, ApJ 544 (2000): L27 (reproduced by permission of the AAS)

Fig. 6.9: (a) M. Malkan, V. Gorjian, and R. Tam, ApJS 117 (1998): 25 (reproduced by permission of the AAS); (b) NASA/UMd/A. Wilson et al.

Fig. 6.10: (a) NASA/STScI/R. van der Marel and J. Gerssen; (b) NASA/CXC/MPE/S. Komossa et al.; (c) both of the above

Fig. 7.1: (a) Palomar Observatory DSS*; (b) D. Neumann et al., AA 400 (2003): 811 (reproduced by permission of Astronomy and Astrophysics)

Fig. 7.2: NASA/CXC/SAO/A. Vikhlinin et al.

Fig. 7.3: N. Sharp/NOAO/AURA/NSF

Fig. 7.4: visible: Palomar Observatory DSS*, x-ray overlay: NASA/CXC/INAF-Brera/G. Trinchieri et al.

Fig. 7.5: (a) UK Schmidt DSS**; (b) NASA/CfA/J. Vrtilek et al.

Fig. 7.6: (a) Palomar Observatory DSS*; (b) NASA/CXC/UCI/A. Lewis et al.

Fig. 7.7: (a) Palomar Observatory DSS*; (b) NASA/CXC/SAO

Fig. 7.8: (a) B. McNamara, ApJ 443 (1995): 77 (reproduced by permission of the AAS); (b) VLA image courtesy of NRAO/AUI and G. Taylor; (c) NASA/CXC/SAO

Fig. 7.9: (a) NASA/CXC/IOA/A. Fabian et al.; (b) NASA/CXC/IOA/A. Fabian et al.

Fig. 7.10: (a) Palomar Observatory DSS*; (b) E. Blanton et al., ApJ 558 (2001): L15 (reproduced by permission of the AAS)

Fig. 7.11: (a) UK Schmidt DSS**; (b) NASA/CXC/Ohio U./B. McNamara et al.

Fig. 7.12: (a) NOAO/AURA/NSF; (b) NASA/IOA/J. Sanders and A. Fabian

Fig. 7.13: (a) Palomar Observatory DSS*; (b) NASA/NMSU/J. Pickney et al.; (c) NASA/SAO/CXC; (d) x-ray: NASA/A. Fabian et al., radio: J. Ge and F. Owen, AJ 105 (1993): 3 (reproduced by permission of the AAS)

Fig. 8.1: NASA/SAO/CXC/M. Markevitch et al.

Fig. 8.2: NASA/CXC/SAO

Fig. 8.3: (a) x-ray: NASA/CXC/
A. Siemiginowska (CfA) and J.
Bechtold (U. Arizona), visible:
NASA/HST/CfA/A. Siemiginowska
et al.; (b) NASA/HST/CfA/
A. Siemiginowska et al.

Fig. 8.4: (a) NASA/PSU; (b) NASA/PSU/
G. Garmire et al.

Fig. 8.5: NASA/WMAP Science Team

Fig. 8.6: (a) NASA/CXC/PSU/
D. Alexander et al.; (b) NASA/JHU/
AUI/R. Giacconi et al.

Fig. 8.7: (a) X. Fan et al., AJ 122 (2001):
2833 (reproduced by permission
of the AAS); (b) NASA/CXC/PSU/
W. Brandt et al.

Fig. 8.8: (a, b) NASA/ESA/A. Koekemoer
(STScI)/M. Dickinson (NOAO) and
the GOODS Team

Figs. 9.1-9.2: Courtesy of ISAS/JAXA

Fig. 9.3: (a) NASA; (b) Courtesy of ESA

Fig. 9.4: UMd/C. Reynolds

* Based on photographic data of the
National Geographic Society–
Palomar Observatory Sky Survey
(NGS-POSS) obtained using the
Oschin Telescope on Palomar
Mountain. The NGS-POSS was funded
by a grant from the National
Geographic Society to the California
Institute of Technology. The plates
were processed into the present com-
pressed digital form with their per-
mission. The Digitized Sky Survey
was produced at the Space Telescope
Science Institute under U.S.
Government grant NAG W-2166.

** Based on photographic data obtained
using The UK Schmidt Telescope.
The UK Schmidt Telescope was oper-
ated by the Royal Observatory
Edinburgh, with funding from the UK
Science and Engineering Research
Council, until 1988 June, and there-
after by the Anglo-Australian
Observatory. Original plate material is
copyright © the Royal Observatory
Edinburgh and the Anglo-Australian
Observatory. The plates were
processed into the present com-
pressed digital form with their per-
mission. The Digitized Sky Survey
was produced at the Space Telescope
Science Institute under U.S.
Government grant NAG W-2166.